My Own Right Time

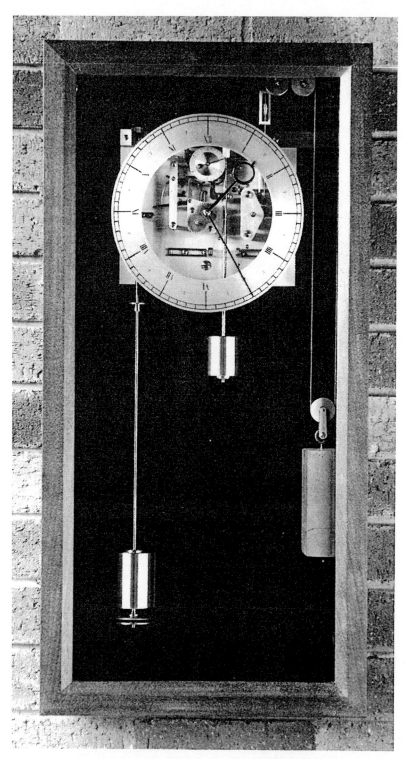

The author's free pendulum clock W5.

My Own Right Time

An Exploration of Clockwork Design

PHILIP WOODWARD

OXFORD UNIVERSITY PRESS

This book has been printed digitally and produced in a standard specification in order to ensure its continuing availability

OXFORD
UNIVERSITY PRESS

Great Clarendon Street, Oxford OX2 6DP

Oxford University Press is a department of the University of Oxford.
It furthers the University's objective of excellence in research, scholarship,
and education by publishing worldwide in

Oxford New York

Auckland Bangkok Buenos Aires Cape Town Chennai
Dar es Salaam Delhi Hong Kong Istanbul Karachi Kolkata
Kuala Lumpur Madrid Melbourne Mexico City Mumbai Nairobi
São Paulo Shanghai Taipei Tokyo Toronto

Oxford is a registered trade mark of Oxford University Press
in the UK and in certain other countries

Published in the United States
by Oxford University Press Inc., New York

© Philip Woodward 1995

The moral rights of the author have been asserted

Database right Oxford University Press (maker)

Reprinted 2003

All rights reserved. No part of this publication may be reproduced,
stored in a retrieval system, or transmitted, in any form or by any means,
without the prior permission in writing of Oxford University Press,
or as expressly permitted by law, or under terms agreed with the appropriate
reprographics rights organization. Enquiries concerning reproduction
outside the scope of the above should be sent to the Rights Department,
Oxford University Press, at the address above

You must not circulate this book in any other binding or cover
and you must impose this same condition on any acquirer

ISBN 0-19-856522-4

Printed in Great Britain by
Antony Rowe Ltd., Eastbourne

Foreword

by Jonathan Betts, Curator of Horology,
Old Royal Observatory, National Maritime Museum,
Greenwich

It is the dream of many horologists to design and build the ultimate mechanical clock, a clock of incomparably good performance and yet of fundamentally simple design, with only a falling weight as a power source and a pendulum, swinging in air, as a controller. The pages of the horological literature are filled with attempts at this ideal, some more successful than others. But if anyone can claim to have approached nearest to realizing such a dream, it is Philip Woodward.

I am especially proud to be asked to write a foreword to *My own right time* because I have been a Woodward fan for many years; to be precise, since 1972 when I first read his description of his new type of grasshopper escapement. I also discovered recently, much to my surprise, that the school clock mentioned in chapter 1 – a wall clock by Vulliamy in fact – was the very same Ipswich School clock which summoned me thirty years later to those very same classrooms.

Philip Woodward is a professional mathematician but an amateur horologist. From a professional horologist this remark may seem belittling. Far from it. Anyone who has even a basic knowledge of the history of horology knows that non-professionals have contributed hugely to the subject: George Airy and William Shortt, for example, both of whom produced state-of-the-art designs which were used with outstanding results in precision clocks at the Royal Observatory. Even the great John Harrison himself, trained as a joiner, might be considered as having been an 'outsider'.

In his preface, the author lists further examples of horological scientists, all of whom made significant contributions to horology. It is no exaggeration to say that Philip Woodward is their natural successor; his clock W5 is, for me, the most beautifully elegant solution to the quest for a high precision mechanical clock, and the story of its development in *My own right time* makes for fascinating reading.

FOREWORD

Anyone seriously attempting to design a clock for accurate timekeeping must first ask himself what we really *mean* by accurate timekeeping. The horological intellectual recognizes that this actually means a clock which is more easily *predictable* in its behaviour. By definition, the time told by each and every working clock is a perfectly accurate response to the natural influences upon it. What we call 'errors' in timekeeping are merely those influences upon a clock which we haven't compensated for in its design, or which we haven't accounted for in interpreting what the clock says. Philip Woodward considers both these concepts carefully, and as well as incorporating compensation into his own clock designs wherever practicable, discusses the whole question of data analysis in a typically thorough and scientific way. This work is both challenging and new, but with helpful analogies a potentially dry subject is made both interesting and accessible.

With a title like *My own right time* we are led to believe that this is a book not to be taken too seriously. Don't be deceived. It is an important and scholarly work which will prove to be a landmark in our horological literature.

Author's preface

The pendulum never ceases to be a source of interest to scientists, yet the mechanisms used to maintain it in oscillation are strangely neglected in the general history of science and technology. In the past, escapements have received attention from scientists of the first rank, such as Galileo, Huygens, Airy and Kelvin. Others, like John Harrison, Pierre Le Roy, Lord Grimthorpe and William Shortt, to name but a few, are almost unknown outside the horological fraternity. Here I must echo the aspirations of Professor David Landes of Harvard University upon discovering the wealth of interest to be found in the history and science of time measurement. In the preface to his splendid book *Revolution in time*, he writes,

To come upon a major aspect of the development of modern society, economy and civilization that is still uncharted territory is a rare strike. People have asked me how it is that so important a subject has been so little studied. I'm not sure that I know the answer; but this I hope: that time measurement will never again be so ignored.

Perhaps because I myself am an amateur in the horological craft, I have felt able to write with other amateurs and newcomers to the subject constantly in mind. The arcane but traditional terms of the clockmaker, such as arbors, detents and foliots, are explained in more familiar language before use, and I have also prepared a brief glossary.

By profession I am a mathematical scientist, but no Stephen Hawking, so I would find it hard to write about pendulums without introducing a few formulae. Even among scientists, books dense with equations have always been unpopular, but the general phobia of simple mathematics is nowadays exaggerated. Also to be considered are those who find mathematics useful. As Professor Sir Roger Penrose has remarked, equations can always be skipped; they may become clearer at a later stage.

The way in which I have been influenced by Rawlings' seminal work, *The science of clocks and watches*, cannot be hidden, but *My own right time* is a more personal story. In

AUTHOR'S PREFACE

something approaching chronological sequence, it relates how my understanding of clockwork has progressed with the passing years. At first one is content to know what makes a clock tick, but soon afterwards one finds oneself delving into a very strange world where timekeeping accuracies of one part in a million are the order of the day. The fascination lies in seeing how this can be done with purely mechanical tools that have not changed in centuries.

My preoccupation with escapements will be obvious. They are a delight to the geometrical mind, and my only regret is that the drawings cannot be made to move. Video animations of escapements, pioneered by John Redfern, are already beginning to appear in museums. For this book, the computer has enabled me to 'operate' some of my diagrams to check that they would actually work. In too many books of old, the drawings would jam as soon as they tried to move! I have not labelled all the wheels with arrows, the convention being that wheels go round clockwise unless otherwise indicated.

Many of my horological friends and acquaintances have contributed to the activities that made this book possible, principally Charles Aked, Douglas Bateman, Peter Baxandall, Peter Brain, Martin Burgess, James Chandler, Geoffrey Goodship, John Griffiths, Professor Roger Irving, Charles Brandram Jones, Andrew King, Henry Marcoolyn, Anthony Randall, Dr Timothy Treffry, and John Warbey, all of whose kindnesses are gratefully acknowledged. For that most important contribution an author can receive, the first reactions of a reader, I was privileged to have those of Jonathan Betts, Curator of Horology at Greenwich, who not only read and commented on the whole book in manuscript but has generously graced it with a foreword.

Between an author's manuscript and the production of a finished book there are many stages, thankfully not all mechanical. To have these handled with the skill and courtesy shown by the staff of the Oxford University Press is something it gives me great pleasure to acknowledge.

Malvern P.M.W.
1994

Contents

1	**A horologist in the making** *From Meccano to radar – a grounding in noise*	1
2	**Theory and practice** *Principles of resonance and the pendulum – I make one*	8
3	**Choosing an escapement** *Gravity escapements – electricity to the rescue – too noisy*	19
4	**Echoes of Hope-Jones** *Conversion to Synchronome – still too noisy*	32
5	**Harrison and Congreve** *The sublime grasshopper – Congreve's half baked ideas*	38
6	**Silence for a cellist** *Exploiting Harrison's hop – a visit by Paul Tortelier*	46
7	**Going without gears** *Le Roy's one-wheeler – a different solution – Crane's daisy wheel*	50
8	**Disturbed harmonic motion** *Escapement theory by sinusoids – the diva and the wine glass*	63

CONTENTS

9	**The phase circle** *Applying Airy's theory – balancing errors – Brocot's escapement*	72
10	**The Shortt free pendulum** *Hope-Jones as backseat driver – climax of the pendulum era*	82
11	**Aiming too high** *173 vibrations in 4 minutes – overloading the slave – clock stops*	92
12	**W5** *I do my best – one second in 100 days – occasionally*	95
13	**Error correction** *Drift and other systematic errors – least squares analysis*	107
14	**Noise modulation** *White noise – random walk – flicker noise – measurement of instability*	118
15	**The enigma of flicker noise** *A mathematical model – why flicker noise does not cancel*	127
16	**Wallman's conjecture** *Noise and chaos – earth tides – gravity fluctuations*	133
17	**Clockwork with a difference** *Continuous motion – William Bond's regulator – Kelvin's free pendulum*	139
	Appendix	150
	Glossary	152
	References	158
	Acknowledgements	161
	Index	162

CHAPTER ONE

A horologist in the making

The detour through the arboretum gave us time to collect a fresh supply of conkers, but we had to judge things carefully if we were not to be late for school. An old gentleman taking his morning walk in the park might possess the time, but he could prove grumpy when accosted by a couple of seven-year-olds in caps. We would put on our humblest expressions and ask for 'the right time, please', a manoeuvre that could be self-defeating as it would take him an age to reach into his pocket, extract the watch and open it up. Gold hunters on chains were the watches we preferred, though it soon became obvious that the time we were given was unlikely to tally with the school clock. Indeed, watches never agreed with one another, and I was becoming obsessed with the desire to have my own 'right time', accurate to a minute. My father had said it could always be obtained at the Post Office, an observation firmly grounded in history but one that seemed to have little relevance to my immediate needs.

It was the time of the great depression, and my father had already traded in his gold watch for a fat and ugly nickel-plated pocket watch called Federal, which cost him 6s 6d. It had a coarse sounding tick and the dial was made of cardboard, but he made it last for years. Under the circumstances I could hardly plead for a watch of my own, but I was instead given a clock kit which went by the name of UMAKA clock, Figure 1.1. This could be bought from a toyshop; Noel C. Ta'Bois (1984) tells us that it was patented in 1922 by one William Frederick Coman. Many an amateur must have learned about clockwork from this instructive product. When the axles, properly called *arbors*, have been set in the right holes and the plates fixed with the nuts provided, the clock is hung from a nail in the wall and driven by a weight hanging on a chain after the fashion of a cuckoo clock. Seeing how the bits and pieces fit together is one thing, but it is another matter to understand why they are designed as they are. I began to think about this quite hard.

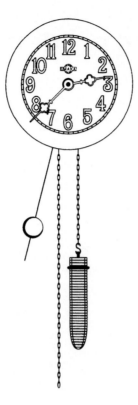

FIGURE 1.1
The UMAKA *clock.*

The UMAKA clock's pendulum is caused to oscillate by *pallets* formed from a bent strip of metal riveted to an arbor to which the pendulum is directly attached, Figure 1.2. The pallets engage with an *escape wheel* of thirty-nine teeth, which would spin round clockwise if the pallets were not there to prevent it. The pallet surfaces at the ends of the bent strip are so shaped that the motion of the escape wheel keeps the pendulum swing-

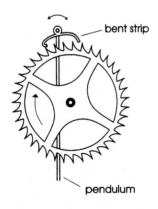
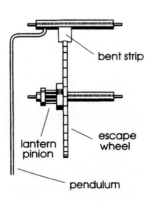

FIGURE 1.2
Escapement of the UMAKA
clock, front and side views.

ing. The chain for the driving weight is draped round a pulley fitted with spikes to engage with the links of the chain. This is not especially interesting and I have not included it in the drawing. The driving pulley is geared to an intermediate arbor carrying a second gear which meshes with a lantern pinion fixed to the escape wheel. The minute hand is driven from the slow moving sprocket arbor, and the hour hand is geared from the minute hand. I use the present tense because, in a house not far from my own, a UMAKA clock is still working well after more than half a century of careful maintenance. My own was cannibalized in the cause of science many years ago!

The UMAKA clock is so simple that one can only marvel that it was not dreamed up in the Middle Ages, but the pendulum, surely the most obvious of mechanical devices with which to beat time, did not figure in anybody's imagination – save that of Leonardo da Vinci – until the middle of the seventeenth century. Galileo had conceived of a pendulum clock before he died in 1642, but it seems certain that the first person actually to build one was Christiaan Huygens, the Dutch mathematician, in 1656. For centuries clocks had used the 'verge and foliot' system, Figure 1.3, in my opinion the most difficult of all escapements to draw! A crown wheel, shaped like a hacksaw blade bent into a circle, operates on pallets fixed to an arbor called the *verge*. The pallets point in roughly perpendicular directions like sawn-off boards on a signpost and engage with the crown wheel at two diametrically opposite points. The verge carries a balance, the so-called *foliot*, which is caused to oscillate in a horizontal plane with no help at all from gravity. In a large clock, the weight of the foliot structure would have introduced considerable friction from endthrust in the lower bearing of the verge, so the weight was taken off it by a string tied to a bracket at the top of the clock. A miniature form of this escapement was used in watches – without the string! – until quite late in the nineteenth century. The balance took the form

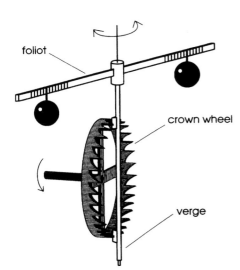

FIGURE 1.3
The verge and foliot escapement of early clocks was driven by an escape wheel in the form of a crown and regulated by sliding the weights towards or away from the centre of the foliot bar.

of a wheel like a hoop in the plane of the watch, whilst the crown wheel had to be accommodated rather awkwardly edgeways on.

My family moved to Cornwall and I was sent away to school, where I did at last have my own pocket watch called Dimrex, a brand name I have never seen since. One day, during a lesson in Thucydides, the winding stem fell out, causing me to fidget in some desperation. My inattention to the Peloponnesian War resulted in confiscation of the watch, but it was returned to me next day with the winder fixed. The moral was clear: if schoolmasters could mend watches, then so could I. Tradesmen have always taken a poor view of do-it-yourself repairers – 'bodgers' – but I would be prepared to debate that they have profited handsomely from enthusiasts such as myself. There are three golden rules for bodgers: know your limitations, do nothing irreversible and never meddle with antiques!

I soon found out how to regulate a watch, though I had not yet discovered how to open the case without risk both to the watch and to my fingers. There is no such thing as privacy in a boarding school, and others were quick to employ my services as an adjuster. I acquired the knack of testing a watch within about ten minutes by holding it against my own and listening to the combination of ticks. In those days, all watches made five ticks to the second, and it was easy to wait until those of one watch coincided with those of another. Equally recognizable was the doubly fast beat of two watches ticking in alternation; I calculated that if it took ten minutes to pass from the one state to the other, the pair of watches differed in their rates by about 15 seconds a day. In half an hour on a Sunday afternoon, I would return a client's watch duly regulated for a fee of one apple, but the promises of accuracy were seldom fulfilled as I had yet to learn about the imperfections of temperature compensation or indeed the total lack of it, of positional errors which caused watches to go at different speeds when 'dial up' or '12 up', and lack of isochronism which made the rate vary as the mainspring ran down. I also began to suspect that my own watch might not be perfect in every one of these respects.

A few years later, I ventured to make a clock from Meccano, using a lump of Cornish granite from the rockery as the driving weight. The escapement was a problem, the urge to invent my own ending in failure. The purpose of an escapement is to convert rotary motion from the train of gears into reciprocating motion to drive a pendulum. Combustion engines have the opposite task, so I tried to reverse the idea as shown in Figure 1.4. The combination of crank and connecting rod seemed simple, but in action it was not a pretty sight; the only way to allow the pendulum any latitude in its arc of swing was to leave the nuts and bolts loose! Furthermore, it consumed a whole revolution of the driving wheel for every double swing of the pendulum. What I did not know at the time was that just such an escapement working on the crank principle had been invented in all seriousness in 1799 by Simon Goodrich, who received an award for it from the Society for the Encouragement of Arts, Manufactures and Commerce (the present-day Royal Society of Arts). This escapement is illustrated in Rees' *Cyclopaedia* (1820). Goodrich engineered his loose connections more cleverly than I had done, using chains and springs. The virtue

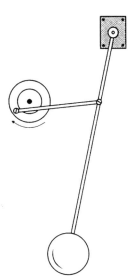

FIGURE 1.4
Woodward's crank escapement.

claimed was silent running, but – as may well be guessed – this was not a noticeable feature of my own version.

Chastened, I copied the escapement of the UMAKA clock and found it to work well enough, but in retrospect the part of my Meccano clock of which I am still rather proud was the chiming arrangement using the principle of tumbling dominoes. Not long ago I published an account of this in the Horological Journal to see whether it had been done before, and to my surprise nobody responded.

To sound the hour, most mechanical clocks strike a gong or bell the appropriate number of times, but it complicates the mechanism greatly and I could not face up to that. There is a much simpler arrangement often used in skeleton clocks known as a 'passing strike', in which the bell is struck just once at every hour. It is no use for telling the time in the night, but it costs very little to make. I thought it would be more ornamental to strike up a little tune; such a 'passing chime' would surely not be too hard to contrive, given a few old bicycle bells.

For thirty minutes before the hour, a bell hammer loosely mounted on the minute hand spindle was gradually lifted to top dead centre by a pin on a crank. A trifle past top dead centre the hammer went over, and on its way down struck a bell. Stacked loosely on the same spindle behind this hammer were the others, all being lifted together about one minute behind the first one. After striking the bell, the first hammer swung to the bottom and struck the tail of the second hammer with sufficient force to send it over the top. After this one had struck its own bell, it jolted the next one, and so on for as many hammers as there were bells. I had only three bells at my disposal, and as luck would have it, they chimed soh, me, doh. It was the last time I ever used Meccano.

World War II burst in on this scene in an unexpected way, for during the very first night in which my clock was allowed to run unattended, I was awakened by a loud bang which

FIGURE 1.5
Blip on an A-scope radar trace.

shook the house. I naturally supposed that the granite boulder had crashed to the floor, but it turned out to be a German bomb exploding in a field on the opposite bank of the Truro river, serving to remind me that horological exploits would soon have to be put away for more serious business.

The more serious business into which I was drafted was to be at the (then) highly secret Telecommunications Research Establishment, where I had the extraordinarily good fortune to work alongside the cream of Britain's young scientists on 'radiolocation', later to be called radar. As a raw graduate I felt like a complete amateur, but after the couple of sterile years I had spent reading mathematics at Oxford, TRE was an eye-opener. Here it was not a matter of *reading* mathematics, but of doing it for real, instructive numerical calculations to start with and theoretical analysis later on. I vividly remember how mathematics sprang to life when it expressed the nub of a physical problem. Unlike some of the Oxford dons who had taught me, I have never felt that mathematics is debased in serving the cause of physics. Most mathematics was developed in response to science, and we need look no further than Newton and Einstein to understand how physics itself advances when its ideas are distilled into a formal mathematical framework.

It was at the nearby operational radar station in a Dorsetshire hamlet with the delightful name of Worth Matravers that I first saw a radar trace on an 'A-scope'. It resembled my computed simulation at Figure 1.5. The tube was scanned from left to right in about one millisecond, starting from the moment a radar pulse was transmitted. Any echoes received were shown, not as dots of varying intensity as on a modern plan-position indicator, but simply as vertical peaks or *blips*, from targets in all directions. A blip on the trace at the extreme right would theoretically represent an echo that had been delayed by a whole millisecond, during which time the pulse would have had time to travel 93 miles to the target and 93 miles back. The operator's duty was to pick out an approaching aircraft as soon as possible and measure its range as distance along the trace, the obvious difficulty being to distinguish genuine blips from electronic noise generated within the receiver. Operators became skilled in reporting targets in a matter-of-fact tone which belied the inherent uncertainty of the task.

One of the more obvious objectives of the radar designer was to maximize the signal-to-noise ratio, which depended crucially on the use of suitable filtering circuits. However, it struck me as being more of a problem in the theory of probability, and some years later I began to follow up the work of that brilliant American mathematician, Claude Shannon, whose theory of communication in the presence of noise had opened new doors in the fields of information technology and computing science. In consequence I found myself in

1956 invited by physicist J. H. Van Vleck to give a course of lectures at Harvard on random processes. By a lucky chance, Shannon had been invited to give a course of lectures 'down the road' at MIT, and these I attended with enthusiasm. The relevance of all this to horology may not be clear, but noise and random processes are not confined to radar and communications. The performance of a precision clock is ultimately limited by random disturbances which fall in the same category as noise. Mathematically they are studied with exactly the same tools.

Today, of course, we are beginning to forget the uncertainties of time that used to beset us before the electronic revolution. Anxious railway passengers would frequently break the reserve of the compartment with the question 'What do you make it?', and the answers could stir up a degree of entertaining controversy. Such a question has now moved behind the scenes to laboratories where even quartz crystal oscillators may not be as accurate for scientific purposes as we would like.

Interestingly, the mathematical theory of time measurement in the presence of noise has advanced more since the advent of the quartz oscillator than in all the years before it, and has owed much to the researches carried out in the field of radar. In an extensive monograph on time and frequency prepared by the US National Bureau of Standards, Allan et al. (1974b, p. 210) wrote as follows on the subject of reading the time from a clock:

If the spectral character of the noise is known, then in principle one could design an optimum filter in some sense to best examine the signal for frequency drift, frequency offset, and time residual through the noise.

There are deeper waters here than I would have dreamed of when I first started taking an interest in horology – waters into which one may still dive with little fear of striking the bottom. This applies especially to a form of noise called *flicker noise*, to which every form of clock is subject. But that is a topic for a later chapter.

CHAPTER TWO

Theory and practice

Early clocks which measured the burning of candles or the flow of water or sand had no chance of rivalling the accuracy to be obtained by counting vibrations. Without any natural rhythm, continuous flow is unlikely to remain constant to closer than about 1%, which is a quarter of an hour a day. The same might be said of the time taken for a child to slide down a banister rail at a speed dependent solely upon friction, which is surely the most variable of all mechanical forces. That eye-catching clock invented by Sir William Congreve, in which a ball rolls down a zig-zag groove in a tilted metal plate, works on much the same principle, though the friction is rolling rather than sliding. By contrast, the beat of a high quality resonator is virtually independent of friction, being determined by the interplay of two more stable qualities, *inertia* and *elasticity*. These are the fundamental characteristics that make all types of vibration possible, from a bowl of jelly to a quartz crystal.

Some readers may need reminding that even electronic watches are governed by purely mechanical vibrations, for quartz crystals reverberate on much the same principle as the wooden bars of a xylophone. The quality of a crystal's resonance, however, is about a million times higher than that of a wooden bar whose sound dies away almost before any musical note can be detected. A musician's tuning fork gives a clearer note than a xylophone and it is no accident that resonating forks should once have been used in watches, for time and pitch are closely related concepts. A fork vibrating at a frequency of 440 hertz sounds the note A, but we might just as well say that one second is the time taken for the prongs to make 440 vibrations. By counting the vibrations, therefore, and dividing by 440, we have a clock that measures seconds. The fork in Bulova's Accutron wrist watch vibrated at 360 Hz, and the vibrations were counted – amazingly – by a mechanical pawl attached to one of the prongs which turned a tiny ratchet wheel with 300 microscopic teeth.

The interplay of inertia and elasticity has been a source of fascination to physicists for centuries. Elasticity is characterized by a restoring force that always tries to restore a system to its restful state, whilst inertia is a property that resists change. When everything is stationary in the restful position nothing happens, but if the system should be disturbed, it oscillates. It is urged back to its resting position by the restoring force, where its inertia prevents it from stopping. Eventually, the restoring force does bring it to rest, but not at its natural resting position, so the force continues to act until that position is reached once more. Again the system overshoots, and so on from one side to the other. With no friction or air resistance to damp out the vibrations, oscillation would continue indefinitely.

Although inertia and elasticity are descriptive terms, they can be quantified in ways suitable for any given type of resonator. Such definitions are vital to the science of resonance because they enable us to calculate the vibrational frequency; the greater the inertia or the weaker the elastic force – or both – the slower are the vibrations.

The usual example of a resonator is a brick bobbing up and down on the end of a coil spring, and in this system, the inertia is the *mass* of the brick, whilst elasticity can be measured by the *compliance* of the spring. Compliance is the amount of 'give' in the spring per unit of force applied, in other words the opposite – more correctly the reciprocal – of stiffness. The longer the spring the greater the compliance; the bigger the brick the greater the mass. Textbooks such as Rawlings (1993) tell us that the period of each vibration for this simple resonator is 2π times the square root of the product of mass and compliance. These, then, are the properties a clockmaker must strive to keep absolutely constant.

Mass does not present any problem, but compliance can be trickier. Some types of spring can exhibit what a laboratory colleague used to call 'silk stocking effect'. At first the stocking stretches elastically, but it reaches a limit beyond which no further stretch is possible. The compliance has progressively diminished to zero. Compliance varying with the state of the system is a highly undesirable quality in an oscillator, as it causes the resonant frequency to vary with the amplitude of vibration, a defect known to watch and clockmakers as *anisochronism*, or unequal times.

The brick suspended on a coil spring is described as a 'lumped' system because nearly all of the mass resides in the brick and all the compliance in the spring; the two properties are in separate lumps. The balance and hairspring resonator in a mechanical watch is also lumped, but a quartz crystal is not. There, inertia and elasticity are distributed throughout the same material, but they determine the resonant frequency in just the same way.

Of all systems, the pendulum is the very worst example to take when explaining resonance, because its elasticity does not come from a physical spring, and the frequency of vibration does not depend on the mass that gives rise to the inertia! To understand this, it is necessary to make a distinction between mass and weight, which can be confusing even to those who once understood it. Mass is an inertial property, but weight is a gravitational force. Today we have only to imagine travelling in space to make the distinction, for in space weight vanishes, while inertia remains unchanged. On earth weight happens to be proportional to mass, and the constant of proportionality is g, which is about 10 in SI units. So accustomed are we to this proportionality that we can easily lose sight of the distinction, and to make matters worse we mix up the terminology. The '2 kilograms' marked on a bag of flour is its mass M, not its weight, for its weight is a force of Mg, about 20 newtons.

The theory of the pendulum is explained in every book on mechanics, and I shall summarize it only to the extent of pointing out why it is so paradoxical. In doing so, I make the simplifying assumption that the angle of swing is very small. The displacement of the bob is $L\alpha$ where L is the length of the rod and α is the angle it makes with the vertical in radians. The restoring force, to a high degree of accuracy, is the weight Mg times α. As this is proportional to the displacement, it behaves just like a spring, and therefore has a compliance. By definition this is the ratio of the displacement to the restoring force, or $L\alpha/(Mg\alpha)$. After cancelling out α, this is $L/(Mg)$. To get the period by the rule given earlier, we first need the product of mass and compliance, which is $ML/(Mg)$, or after cancelling out M, just L/g. The mass has disappeared because the weight is proportional to it. The final result for the period, after taking the square root and multiplying by 2π, is the familiar formula $2\pi\sqrt{L/g}$.

This piece of theory is essential when one tries to understand the barometric error of pendulums, a tiny effect which can slow down an ordinary pendulum by upwards of a hundredth of a second a day for every millibar rise in air pressure. The rise in pressure increases the density of the air and makes the weight of the bob slightly less by Archimedes' principle. This is known as the *flotation* contribution. In addition, the effective mass of the bob increases slightly when the density of the air increases, because more air is set in motion by the bob, and its inertia has to be added to that of the pendulum. These two effects, flotation and *accession to inertia* – as it is known to designers of loudspeakers – reinforce one another and upset the simple relation between weight and mass. The upshot is that the clock loses in high pressure. Extra air pressure also increases air resistance and reduces the amplitude of swing slightly, which introduces another small error if the pendulum is not isochronous.

Many centuries ago, all these ideas of resonance would have been quite foreign to clockmakers, whose attention was focused almost exclusively on the driving weights and gears. The resonator, if such it could be called, was a horizontal bar or foliot (see Figure 1.3) suspended at its centre and capable of swinging aimlessly like a compass needle without any magnetism. There was no intrinsic restoring force; it was up to the escapement to push the foliot one way and then the other. There is historical evidence that the escapement was regarded simply as a brake to keep the clock running as long as possible before it needed rewinding. Physicists may marvel at the fact that such a system persisted for so long before being superseded by the pendulum, but it worked and kept time well enough for the purposes of a less hurried age.

Once Galileo had experimented with the pendulum and Huygens had introduced it to clockmakers in the mid-seventeenth century, horology had entered the modern age. The resonator was at last recognized as the source of accurate timekeeping and the rest of the mechanism as subservient to its needs. For any understanding of horology, this distinction is essential. The task of the 'clockwork' – as I like to call it – is twofold: to count the vibrations and so tell the time, and to counteract frictional forces in the resonator which would otherwise damp out the vibrations. In carrying out these tasks, clockwork necessarily interferes with the vibrations and in so doing destroys some of the intrinsic timekeeping accuracy. That is where the simple theory of vibrations ends and the theory of oscillation begins.

Looked at from a theoretical point of view, the horological breakthrough initiated by Galileo centres on the restoring force. Inasmuch as the old foliot did actually vibrate, the force that drove it had something of the quality of a restoring force, but it was by no means proportional to the displacement of the mass. The arrangement therefore lacked isochronism; the frequency of vibration depended on the amplitude of swing. The amplitude depended on the strength of the drive, which might vary for any number of reasons. In a watch driven by a mainspring, timekeeping varied grossly as the force of the mainspring diminished through the day. The worst of this trouble was cured by the introduction of the balance spring in 1675. Like the pendulum in a clock, this supplied a constant compliance independent of the drive, whose contribution then became less important. Ideally, the driving force should do no more than counteract friction in the resonator without adding any restoring force of its own. Its purpose had completely changed.

Cleaning up the driving force is what precision clockwork is all about; many and various are the interesting methods that have been tried. One of the first to understand these matters deeply was John Harrison, who stated that the pendulum must have 'dominion' over the clockwork. The less friction there is in the resonator, the less work the escapement has to do to overcome it and the less it interferes with the timekeeping. Harrison invented a virtually frictionless escapement (described in another chapter) and was painstaking in finding ways to keep the driving force constant.

If friction in the resonator – from whatever source – is small compared with the restoring force, the resonator will continue vibrating for many cycles without being driven at all, like a bell resounding long after being struck. That is the test for what may be called the *quality* of resonance, and the symbol Q has been adopted as a numerical measure of it. Q is the ratio of the peak restoring force to the peak damping force. It is a dimensionless number that originated in the field of electrical circuits, not mechanical horology, and for some reason horologists did not at first take kindly to it. Be that as it may, Q is a useful ratio with a clearly defined meaning, and one having an obvious connexion with accuracy of timekeeping, as Jespersen and Fitz-Randolph (1982, chapter 4) make abundantly clear in their popular account of time and frequency. The history of Q has been documented of late by several writers, though none so beautifully as Estill Green (1955).

The Q of the balance and hairspring in a mechanical watch might be no more than 100, whilst that of a good pendulum in air might be 10 000. (In a vacuum, it can be 100 000.) The Q of a quartz crystal can be 1 000 000 and that of the resonator in an atomic clock can be 100 000 000. Leaping as they do by factors of 100, these figures alone go far to support the argument that a high value of Q is the secret of accurate timekeeping, though it might be less contentious to say that a high Q is just one of the necessary conditions. In any one régime, such as that of pendulums, all kinds of other things matter as much or even more.

Theory is one thing, beautiful in its abstraction. Practice is quite another, as I was now to find out for myself when I embarked on making my first seconds pendulum. It was nothing to be proud of, and I tell the tale if only to entertain those amateurs who have gone through similar experiences themselves. As a theoretician, I naturally wanted the Q to be as large as possible. The energy of a pendulum is lost partly to the air through which it swings, partly in friction (or its equivalent) at the point of suspension, partly in the reaction of the whole supporting structure and partly to the very mechanism that supplies the pendulum with energy, which may itself be lossy.

To reduce air resistance, one would ideally like all the mass of the bob concentrated at a point. Substances of infinite density are not available on this earth, so I would have to make do with lead. Tungsten is 1.7 times denser, but even had I thought of it, it would have been ruled out on grounds of cost and difficulty of working. It is just possible to cast lead in the kitchen at home, though it is not a task I enjoy and is not something I would recommend or wish to repeat. Apart from the noxious nature of the metal the thought of an accident is frightening, for the handle of an ordinary saucepan may easily collapse under the weight. As for the bob shape, I chose a circular cylinder simply because I possessed a suitable can to use as a mould. When the cast had cooled, the tinplate peeled off cleanly like the lid of a sardine tin, possibly because it had once held motor oil. It was beginner's luck, for whenever I have tried casting lead since that well remembered day, I have found tinplate and lead to be quite inseparable.

Although many precision clocks have cylindrical bobs, Bateman has shown it to be about the worst possible shape for low air resistance (Rawlings 1993, pp. 89–94). A sphere is considerably better. I had not forgotten yet another way of eliminating air resistance, which is to run the clock in a vacuum, as was the practice for certain observatory clocks. In choosing to ignore the existence of tungsten or the possibility of a vacuum, not to mention the problem of casting a sphere or attempting to machine it in my tiny lathe, my principles were now so compromised that it hardly seemed to matter what I did next, and there were still the suspension and the support to consider.

The support I left until last. It should perhaps have been considered first, for as the heavy pendulum swings from side to side, it tries to rock the clock, or even the house. This in itself would not matter if everything were perfectly rigid, or even perfectly elastic, but such is not the case and vibrational energy is lost as a result. There are only two ways of combating this, one of which is to use a double pendulum, with two bobs swinging in contrary motion, and the other is to use a lighter pendulum on the grounds that a small enough tail cannot wag the dog. Unfortunately, I had chosen the tail already and had now to attach it to the dog using stout cast iron brackets. This seemed an impossible task for a theoretician until I remembered having kept the solid iron lugs from an old bedstead. Filed to shape, they served admirably. They were bolted to a machined iron plate a quarter of an inch thick recovered from a scrap heap, and this in turn was bolted to a hardwood backboard one inch thick and screwed to an outside wall of the house. A friend more practical than myself told me the structure would safely have suspended a family saloon car. This I took as a compliment, for heavy engineering is not my forte, but I dislike brute force solutions.

Three centuries ago, Isaac Newton (1687) slipped up with his pendulum support. He had been doing some experiments on air resistance, his pendulum bob consisting of a wooden box of weights, suspended from a steel hook by a thread eleven feet long. He wrote,

The first time I made it, the hook being weak, the full box was retarded sooner. The cause I found to be, that the hook was not strong enough to bear the weight of the box; so that, as it oscillated to and fro, the hook was bent sometimes this way and sometimes that way. I therefore procured a hook of sufficient strength, so that the point of suspension might remain unmoved, and then all things happened as is above described.

For my pendulum rod, I chose invar, an alloy which expands and contracts very little with changes of temperature. The length of the rod is what determines the frequency of vibration, so the more stable its length the better. At first, invar sounds ideal. The coefficient of expansion varies from one sample to another, but is so small that my Sheffield supplier could not give me any significant figure for it. It is less than one part per million per degree Celsius, compared with about eleven parts per million for steel. Ignoring every other factor, a pendulum rod which expands by one part in a million per degree will cause a clock to lose a little over 4 seconds in 100 days per degree C rise in temperature. This compares with 48 seconds for a rod of ordinary steel. It shows the virtue of invar, but invar is a peculiar material about which there are some vexed questions.

Metallurgists tell us that invar is an unstable alloy whose constituents are under continuous internal stress, as a result of which it is apt to fidget and change its length spontaneously, regardless of temperature. Such changes may be less than one part in a million, but they are big enough to limit the performance of a precision clock. Invar is supposed to settle down if allowed to rest for some years before use, but my rod was new and I needed to use it straight away. It was described as 'cold drawn', which, if I understand the term correctly, means that it has been pulled through a hole as though it were made of chewing gum. Such brutal treatment must surely call for careful annealing, but no such treatment had been applied to my rod by the maker, and I had to take it or leave it. All one's scientific instincts rebel at not having things under proper control, and here was yet another compromise.

By now I was beginning to look for the easiest solution to every problem, and had no qualms about supporting the lead bob by its base, finished tidily with a disc of brass resting

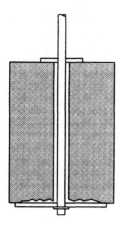

FIGURE 2.1
Cross section of a pendulum bob, roughly cast in lead, cylindrical in shape and with a central hole. The rod is located by discs fixed to the top and bottom surfaces of the lead, and the whole is supported by a pin through the rod.

on a pin through the rod, Figure 2.1. This form of construction causes the bob to expand upwards in heat, tending to make the clock gain slightly. The upwards expansion of the centre of gravity of the bob can be avoided by supporting the bob at its centre, but this entails forming a shoulder inside it. If any compensation for the tiny expansion of the rod is needed, it is traditionally supplied by interposing a short sleeve of brass or aluminium between the shoulder in the bob and the support point on the rod, the length of the sleeve being found by experiment. I had decided to forget all about trying to form a shoulder inside a lump of lead. A different solution offered itself when I came to make the suspension.

The pendulum of a good clock is never suspended on an axle like that of the UMAKA clock, because the friction in its bearings would reduce the Q intolerably. The choice is between knife-edges and springs. In a laboratory 'string-and-sealing-wax' experiment, I have seen a light pendulum supported by the edge of a razor blade rocking on a sheet of glass, which sounds delightfully simple. However, when this method is suitably refined, it is no easier and scarcely any better than the more usual short ribbon of spring steel, firmly clamped to the support at its upper end and to the pendulum rod at its lower end.

The fixing of the suspension spring is not quite as straightforward as it may sound. My form of construction, hardly original, was to bolt the ribbon between a pair of metal blocks at the top, with a pin protruding on each side to rest in V-grooves in the support, Figure 2.2. The lower end of the spring was likewise clamped, and the sandwich supplied with a pin protruding on each side, as at the top. Two parallel hooks at the top of the pendulum rod then completed the suspension. Notice that the pendulum is free to swing in two different planes, side to side with the spring flexing, or in and out with the upper pin turning like an axle in the V-grooves of the support. This latter degree of freedom helps to ensure that the pendulum hangs true, but once the pendulum has found its vertical, any swinging in and out is quickly damped by friction.

As may have been guessed from the drawing, I made the pair of hooks for the top of the pendulum rod from two strips of brass between which the rod could slide. The strips are

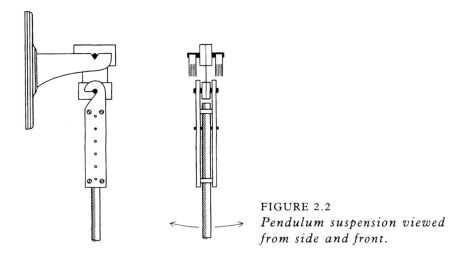

FIGURE 2.2
Pendulum suspension viewed from side and front.

separated by little pillars to form a cage around the rod, after which one pin through the cage and the rod finished the job. Here at last was an opportunity to deal with the problem of temperature compensation. If the strips are long enough, they can be pinned to the rod some distance down from the top. The working rod is now brass for part of the way down, and invar below that. The downwards expansion of the brass can be made to compensate for the upward expansion of the lead bob.

As the expansion of the invar was unknown, the pinning point was to have been found by experiment. Indeed, I have never succeeded in adjusting any form of temperature compensation other than by experiment, for there is much that can vary. The compliance of the suspension spring varies with temperature, as also does the density of the air through which the bob swings. To compute the different effects theoretically is an instructive exercise, but uncertain in the end. The experimental method is not without its difficulties either. The experiments cannot be carried out until the clock is finished and running, so the compensation has to be adjustable. This would be straightforward with the form of construction I had chosen. With a column of holes through the hooks and the rod, the best pinning point could be found by trial and error without altering the period of swing, but I never got round to finishing the job. John Harrison had used this method of adjustment for his famous gridiron pendulum early in the eighteenth century, but in my ignorance of history I was unaware of the fact. I do believe that I would have reinvented the wheel out of laziness had I been clever enough.

The mechanics of adjusting temperature compensation is one thing, but measuring the going rate of a clock in different temperatures is quite another. I soon learned that it is futile to try heating up the room for a short period. Natural variations from day to night are better, but for accuracy one requires a longer period of constant temperature, long enough for the pendulum to have reached thermal equilibrium with its surroundings, and for the rate of the clock to be measured with reasonable precision. My best experiments of this type have been made in August, when there may be a heat wave lasting for a few days to compare with the usual English cool beforehand and afterwards. Changes of temperature are, in my opinion, the biggest enemy of precision timekeeping. After years of observations on another clock, I have formed the impression that any sudden change of temperature can trigger an irreversible change in the length of an invar pendulum rod. The alloy being unstable, any mechanical or thermal shock seems a good excuse for the molecules to rearrange themselves, which is why it is wise (and clever) to repeat measurements made before the heat wave starts with some more afterwards. It is hardly surprising that observatory pendulum clocks, in spite of having invar rods, were always supposed to be run at a constant temperature.

Pendulums do not wear out, and the one I have described could last indefinitely. Had I been more skilled, I might have encased the lead bob in some nobler metal for the sake of appearance, but I am still not sure whether I would have tried to support the bob at its centre. The argument in favour is that the bob, with its high thermal inertia, is removed from the chain of expansions and contractions in heat. The bob answers to a change of temperature more slowly than the other components of the pendulum, so there will be transient losses and gains during heating and cooling. I am doubtful as to whether these are of any great consequence, as the loss caused by the bob's thermal sluggishness during a rise of temperature is cancelled by the corresponding gain during the subsequent fall.

To match the precision of the most accurate mechanical clocks, a pendulum's length should remain constant to within a wavelength of light, yet every junction of materials can introduce uncertainties greater than this. I was probably expecting too much of a structure in which the bob rested on a pin supported by a hole through an invar rod, whose upper end was supported by a pin which rested in a hole through the brass cage at the top, whose side members were hooked over a pin through the suspension block within which the suspension springs were bolted. The length would be more stable if the whole structure could have been carved seamlessly out of a solid bar of metal, with a hole through it to take a knife edge for suspension. I understand that gravimeters used in oil prospecting have been made in just such a way.

A pendulum can be roughly tuned to the desired frequency, in my case 0.5 hertz, even before the clock that will drive it has been made. It can swing by itself for an hour or more, and can be timed to within a second by counting swings – not as difficult as it sounds. How the final tuning is done is a detail. I pinned the rod safely too low at first, timed it, and after careful calculation pinned it again higher up, leaving room for another stage when there was a clock to keep it going. As is usual for precision clocks, fine regulation is carried out by adding weights to a small tray fixed on the rod. If the tray is fixed one third of the way down the rod, Rawlings' rule tells us that a weight equal to one ten thousandth part of the weight of the pendulum will cause a gain of about one second a day.

Many clockmakers will be surprised, and indeed critical, of my choice of supporting the bob on a pin rather than a 'rating nut'. Obviously a threaded nut would make a coarse adjustment to the required length relatively straightforward, but a screw does not strike me as a clean-cut ending for a pendulum. The pendulum is supposed to be stable. It should be made the correct length and stay there!

The first experiment I wished to carry out was to measure its Q. For this purpose, the definition of peak restoring force to peak damping force is not the easiest one to use. A mathematically equivalent definition is

$$2\pi \times \frac{\text{total vibrational energy}}{\text{energy lost per period}},$$

from which it can be proved that Q is $\pi/\log_e 2$ times the number of periods (*double* swings) the pendulum makes in decaying to half amplitude. Using a pocket scientific calculator, $\pi/\log_e 2$ is easily found to be about 4.53. I therefore set the pendulum swinging with a healthy arc of about 1.7° a side and waited for it to decay.

Holding a tape measure behind the rod just above the bob, I measured the total swing – twice the mathematical 'amplitude' – in sixteenths of an inch and called it y. As only the ratios of these measurements matter, the position down the rod is immaterial (provided it remains the same), and so are the units of measurement. Unscientific they may be, but sixteenths of an inch were convenient, and I measured y every five minutes or so. After 45 minutes, the amplitude of swing had only decayed to about 1.2°, and I decided to make that do.

The actual times of each measurement had been noted to the nearest half minute, and the common logarithm of y was plotted against the time in minutes, Figure 2.3. The advantage of the logarithm is that normal *exponential decay* gives rise to a straight line graph, which is precisely what I found. Obtaining a figure for Q was now straightforward

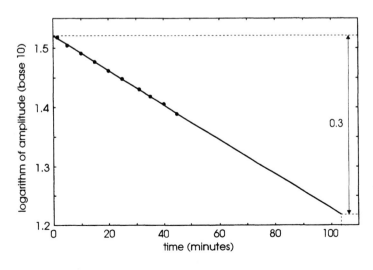

FIGURE 2.3
Determination of Q from a graph of amplitude decay. With a seconds pendulum, 103 minutes is 3090 periods, and Q is then $3090 \times 4.53 = 14\,000$ (approx).

even though I had not waited for the pendulum to reach half amplitude, which is approximately 0.3 down the log scale, 0.3 being the common logarithm of 2. The straight line can simply be extrapolated to the required point, and as may be seen from the graph, it led to a Q value of 14 000. This is not a bad figure for a domestic regulator clock, bearing in mind that the cylindrical bob shape is the least aerodynamic of all.

CHAPTER THREE

Choosing an escapement

The number of different ways of maintaining a pendulum in oscillation is legion, and over the centuries many schemes have been tried – so many, indeed, that it would hardly be practicable even to list them, still less to place them in an order of merit, so I shall describe only those that are relevant to my story. First there are one or two matters of terminology.

It is in the nature of mechanics that a pendulum should be maintained by a succession of little pushes rather than a continuous drive. A force acting through a given distance supplies a pulse of *energy*, known in mechanical horology as an *impulse*, which is unfortunate. To the academic mathematician, an impulse is defined as a force acting for a certain length of *time*, giving rise to a change of *momentum*, which is a different concept altogether. We must simply put up with this misuse of terms, which is hallowed by tradition. Also, it is convenient to generalise the term 'escapement' to cover any kind of mechanism which gives impulse to a pendulum, even though in some cases nothing may actually escape!

For those who would make a clock, choosing an escapement is not easy. Comparative tests of timekeeping accuracy are virtually non-existent, partly because of the difficulty of isolating the effect of an escapement from everything else that can vary. Even when carried out scientifically, such tests are apt to be inconclusive, because there is always room for debate about the statistical criteria that should be applied. We must also remember that accuracy is by no means everything; in everyday life, minimal maintenance is just as important. As so often happens with a technical product, there is one set of variables for the design process and another set concerned with the final suitability to purpose. How the second set depends on the first may be so complex that the optimum solution cannot be computed, but must simply be allowed to evolve.

It is hardly surprising that in situations like this some rather arbitrary criteria should have been put forward. It is often argued, for example, that a pendulum should be given its

FIGURE 3.1
The Riefler pendulum is suspended by a spring anchored on a knife-edge, shown here as a roller. The escapement acts by rocking the suspension.

impulse near the top of the rod, but this is usually a matter of practical convenience, and I would not myself care to say more. The famous maker Riefler, however, made it a matter of principle and took it to its extreme, for the Riefler escapement applies impulse by rocking the upper block of the suspension spring on a knife-edge, shown in exaggerated form in Figure 3.1 as a roller. The extra torque on the suspension spring is so phased that it adds to the energy of the pendulum. The argument against impulsing at some distance down the rod – or even at the bottom – is that any small unwanted variations in the applied force will have a greater effect because of their greater leverage. In his book on the history of the family firm, Dieter Riefler (1981) states this unequivocally, but as the same consideration applies to the impulse as a whole, the argument appears to me fallacious. If the unwanted variations are in scale with the total force, it cannot matter how or where the force is applied.

One counter-argument is that unwanted fluctuations are never in scale with the total force, but are proportionately larger for smaller forces. This is because very small forces seem to be less manageable than larger ones. A large force acting over a small distance is preferable to a small force over a large distance, but this still does not prove that impulsing near the top of the rod is preferable, unless the distance through which the force acts is a fixed proportion of the arc of swing, for which there is no necessity. In any case, account should surely be taken of where the unwanted variations originate. I see no virtue at all in applying impulse through the suspension spring if the variations can be traced to an escape wheel placed at a conventional distance from the axis of rotation of the pendulum. It would then be better to deal with the problem at its source, as Harrison did with the grasshopper escapement to be described in chapter 5. Such is the nature of much horological debate.

Leaving all niceties aside, there are broadly two classes of escapement – those which give impulse centrally in the swing and those which start with a recoiling force. The distinction is easily made if we imagine ourselves trying to maintain a pendulum manually. We might, for example, watch the swings carefully and use our fingers to apply a gentle

CHOOSING AN ESCAPEMENT

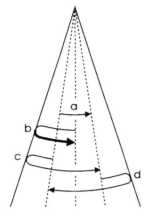

FIGURE 3.2
Types of impulse. (a) At centre of swing. (b) Recoil with crescendo. (c) Recoil with follow-through. (d) The same on following swing.

jab to the rod in mid-swing, possibly a centred impulse as shown at Figure 3.2a. With a slowly moving pendulum this is not difficult, but it does involve watching. Traditional clocks have no eyes, and central impulsing therefore raises some tricky mechanical problems. Forced to work in the dark, the simplest manual method of giving impulse would be to place a finger limply in the path of the pendulum rod, wait for contact, allow the finger to be pushed back by the inertia of the pendulum, and when the pendulum swings forward again, to give it a hearty push, as shown at Figure 3.2b.

If the force applied is greater on the forward stroke than during the recoil, energy will have been imparted, but I know of only one escapement that works in that way – Nicholson's – to be described later in this chapter. Ordinarily, the force has much the same strength in both directions, so the finger must continue pushing beyond the point at which the pendulum first contacted it, as shown at Figure 3.2c, after which the finger must be moved back to its starting position ready for the next vibration.

It is traditional to deliver impulses on every swing, that is twice in each period. This is called *double-beat* action, as at c and d in Figure 3.2 for recoil impulses. The Riefler escapement supplies impulses of this type, as also does the escapement of Big Ben. Double-beat escapements are far more widespread than single-beat escapements, though some electrical clocks cause impulses to be applied at much less frequent intervals, such as half a minute, or perhaps only when the amplitude of swing has fallen to some pre-arranged level, as in the electrically maintained clock designed by Dr Matthäus Hipp of Neuchâtel. An advantage of widely spaced impulses is the larger and more manageable force which can then be used.

The two commonest escapements are the *anchor*, with its recoiling impulse, and the later *dead-beat*, supplying a non-recoiling impulse which is nearly but not quite central. The anchor escapement, so called from the shape of the pallet frame, Figure 3.3, was a natural historical development to suit the first seconds pendulums with their relatively small arcs of swing. It is not known who invented it, but it was used in the turret clock made by William Clement as early as 1671 for King's College Cambridge, and even earlier in the chapel clock at Wadham College Oxford. The anchor escapement of this

FIGURE 3.3
One form of the anchor escapement.

clock is thought to be original, almost certainly the work of the celebrated maker Joseph Knibb (Baillie et al. 1982), whose younger brother John was paid £1 a year by the college to maintain it from 1670 until 1721 (Beeson 1957). It is said that Wren, one of Wadham's Fellows, had a hand in its design. I feel a special sense of attachment to this particular

Movement of a turret clock made in 1670 by Joseph Knibb for Wadham College Oxford, with what is perhaps the earliest surviving anchor escapement.

CHOOSING AN ESCAPEMENT

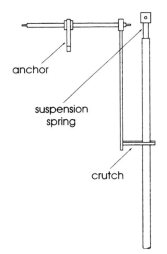

FIGURE 3.4
Connexion between the anchor pallet frame and the separately suspended pendulum in a conventional clock.

clock, which had served the college for two hundred years but which, throughout my own period of residence there, stood as a relic on the floor of the chapel. At that time I had no idea of its historical significance but would give it a friendly twiddle as I passed on my way to the organ loft. It is now in the safe keeping of the University's Museum of the History of Science.

The pallet arbor of the anchor escapement is pivoted in line with the effective point of suspension of the pendulum. Conceptually, it could be rigidly attached to the top of the pendulum, for when the pendulum swings, the pallet frame must swing with it. In practice, the pallet frame and the pendulum are almost always mounted independently (though not in the UMAKA clock). The pivoted arbor used for the pallets will not do for a heavy pendulum, as the friction would be excessive and so would the wear. The arbor for the pallet frame therefore protrudes at the back of the clock frame and carries a dangling arm ending in a crutch, Figure 3.4. This slots in with the separately suspended pendulum rod without having to take any of the pendulum's weight.

The anchor is driven each way by the escape wheel, the final and fastest running wheel of the clock's gear train. The two lateral arms of the anchor are known as *pallets* or pallet arms, and the sloping surfaces at their tips are the business ends, the pallets proper. As the escape wheel tries to rotate, it is obstructed first by one pallet and then the other. The teeth succeed in escaping one by one, and in pushing the pallets out of the way they give impulse to the pendulum. The escapement of the toy UMAKA clock described in chapter 1 is a cheapened version of the anchor.

Let us look at this more carefully. As the acting tooth runs off the end of one pallet, a different tooth drops on to the other pallet and immediately tries to reverse the anchor's motion. The reversal does not take place at once because a pendulum cannot stop dead in its tracks; it continues under its own inertia, taking the anchor with it, and returns in its own good time. Meanwhile the escape wheel tooth is being pushed backwards by the pallet, whose design must leave room for this extra motion. On an old grandfather clock, the

FIGURE 3.5
*The dead-beat
escapement named
after George Graham.*

recoil can be seen clearly by watching the motion of the seconds hand attached to the same arbor as the escape wheel.

By changing the shape of the escape wheel's teeth and pallets, recoil can be completely eliminated as shown in Figure 3.5. This 'dead-beat' escapement has been used with great success since the days of George Graham, who brought it to perfection in 1715, but it is now known (Howse and Hutchinson 1971) to have been made at least forty years previously by Thomas Tompion and by Richard Towneley. Graham had been Tompion's favoured assistant, later his partner and eventually his business successor. In the diagram, a tooth is running along the sloping surface of the pallet on the left and giving impulse as the pendulum swings towards the left. When it drops off, a tooth on the right falls, not on to the sloping surface of the pallet, but on to a 'dead' surface concentric with the pallet arbor. Here the tooth is forced to remain at rest until the pendulum completes its swing to the left and returns towards the centre. As soon as the tooth leaves the dead surface for the sloping tip of the pallet, the wheel starts to turn and give impulse.

This escapement delivers an approximately central impulse, though when one tries to lay out the geometry, one finds that exact centrality is mathematically impossible. For safe landings on the dead faces, the geometry requires the impulsing to be slightly late, the middle of each impulse coming just after the pendulum passes through centre each way. Because of the continuous frictional contact between the pallets and the escape wheel (except during the brief moment of the 'drop'), Graham's escapement cannot be accounted ideal. The *friction rest* between impulses serves no useful purpose in itself and lowers the effective Q of the pendulum. Perfectionists – which horology has never lacked – try to avoid all useless friction, especially that which affects the resonating element, however indirectly.

Because of the motion of the escape wheel, the seconds hand of a clock with a dead-beat escapement pauses at each mark on the dial before progressing to the next with a stately tread. By comparison, the recoiling seconds hand of the anchor escapement can seem a little fussy. It is said that the dead-beat escapement took over from the anchor in high-grade clockwork because it made for better timekeeping, but I would not like to

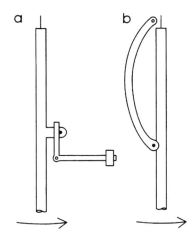

FIGURE 3.6
Two forms of gravity arm operating on the upper part of the pendulum rod, (a) as used in Shepherd's electrical clock at Greenwich, and (b) as used in several mechanical designs. When pivoted close to the pendulum suspension, friction at the impulse pin is minimized.

volunteer scientific reasons for the improvement. Things are often more subtle than they seem. The more dramatic improvement in watches when the recoiling verge escapement gave place to various non-recoiling escapements is much more easily explained. A watch is driven by a mainspring whose force varies as it runs down, unless compensated by an expensive fusee. The force that reaches the resonator during the recoil increases the restoring force and causes the watch to gain, an effect known as *escapement error*. With a non-recoiling impulse, the escapement error can in theory be made zero. The constancy of drive needed for a recoil escapement was not easily achieved in watches, but weight driven clocks enjoy a much steadier driving force, and the argument against a recoil escapement is then correspondingly weaker.

In the interest of learning, I wanted to try something as far away from the traditional anchor and dead-beat escapements as possible. I had seen all too many grandfather clocks with grossly worn pallets, and being keen to avoid the friction that caused it, I decided to try my hand at a *gravity escapement*, the idea of which is simple in principle. To supply impulse, an independent arm presses against the pendulum rod under nothing but its own weight. On completion of the impulse, some suitably contrived mechanism lifts the arm away from the pendulum and holds or latches it up to await its release for the next impulse. One advantage of this scheme is constancy of impulse and another is the almost complete absence of friction during its delivery, both important considerations. In a single-beat version there is a single gravity arm; in a double-beat version there are arms on each side of the pendulum.

A gravity arm can take any number of shapes, two of which are shown diagrammatically in Figure 3.6. The simplest way to reset the arm is by means of an electromagnet, which was first applied to this purpose by Charles Shepherd in 1850 or thereabouts for a clock at Greenwich Observatory. This clock controlled Britain's Standard Time for half a century (Howse 1980) and can still be seen working at Greenwich. I myself decided to use a simpler arrangement devised by Froment in 1854 and later re-invented by many others (Hope-Jones 1949, p. 78).

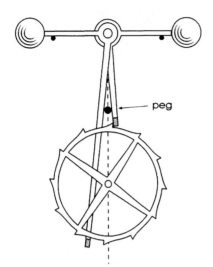

FIGURE 3.7
Nicholson's pseudo-gravity escapement. The broken line indicates the line of the pendulum rod, from which there protrudes a peg to receive impulses.

Electricity in clocks was still a novelty in 1854, but gravity arms were by no means new. Their first appearance on the horological scene is usually credited to Alexander Cumming or (see Cuss 1965) to Thomas Mudge, who may have had the idea from work he did for the astronomer Johann Jakob Huber in 1755. The simplicity of the electrical solution is best appreciated in the context of the somewhat tricky clockwork versions in which the gravity arms are reset by the action of the escape wheel, culminating in Lord Grimthorpe's highly successful design for Big Ben. Here I shall describe a pseudo-gravity escapement devised by W. Nicholson in 1784, then a single-beat variant of Grimthorpe's escapement, and finally a modern gravity escapement by James Arnfield.

The interesting feature of Nicholson's escapement (Rees 1820) is that the gravity arms start and finish their operations at exactly the same place. In Figure 3.7, the escape wheel is trying to turn clockwise but is prevented from doing so by a pallet at the top of the wheel. This pallet is held down by the force of the right-hand gravity arm to which it is rigidly connected, and which is resting on its banking stop. The pendulum is not shown, but its pivot point is in line with that of the two independent gravity arms, and can be imagined to be swinging from left to right through its central position. The pendulum rod carries a pin or peg, seen wedged between the two pallet arms. As the pendulum swings from the centre to the right, it raises the pallet arm, so lifting the weight, and is assisted in doing so by the torque from the escape wheel until the tooth escapes. When this happens, another tooth is immediately detained by the other pallet. Meanwhile the pendulum continues its excursion and in due course returns to the centre accompanied by the gravity arm and pallet alone. In total, the weight of the gravity arm has contributed nothing to the pendulum's energy; the impulse came indirectly from the escape wheel.

Nicholson's escapement is a kind of half-way house towards a true gravity escapement because the clockwork *assists* in the raising of the gravity arm; in a true gravity escapement it takes over the entire task. As so often with half measures, Nicholson has the worst of

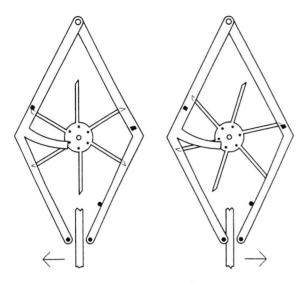

FIGURE 3.8
Thwaites and Reed's variant of the Grimthorpe gravity escapement.

both worlds. Because of pivot friction, the gravity arms cannot even return all of the energy they absorb. They do nothing to ensure constancy of impulse and seem to serve no useful purpose, and yet I find the escapement paradoxically attractive.

Gravity escapements reached their zenith with the design adopted by Lord Grimthorpe for Big Ben, which dates from the same year as Froment's electrical escapement and has been well described by the inventor himself (Grimthorpe 1903). Many variations are possible: Big Ben uses a double three-legged version, but Figure 3.8 shows a single six-legged version (Britten 1938, p. 200, Gazeley 1980, pp. 95–7) designed by Thwaites and Reed. A 'leg' is simply an escape wheel tooth that has grown beyond dental proportions.

Unlike the double three-legged escapement, the six-legged variant is a single-beat escapement. There are two arms hanging loosely from a common pivot positioned as close as possible to the point of suspension of the pendulum. Both arms are gravity arms, in the sense that they are acted upon by gravity (and would tend to cross over if they were free to swing). The arm on the left is the arm that impulses the pendulum, whilst that on the right is neutral. The blobs at the bottom represent pins at right angles to the paper, one on each side of the pendulum rod. These pins are analogous to a forked crutch, except that they are able to move independently. The impulse arm on the left is propped up by one of the six pins of the escape wheel. This in turn cannot move because one of its legs is up against a block fixed to the same arm. (With artistic licence, the arms are shown as semi-transparent.) The neutral arm on the right is propped up by a stop fixed to the frame of the clock. In the first diagram, the pendulum is swinging freely through its centre to the left, and will soon start lifting the impulse arm. As it does so, the escape wheel is released and starts to turn, but it does not get very far before another leg runs into an obstruction on the neutral arm.

By now, something important has happened; the pin which had been propping up the impulse arm has moved round just far enough to be cleared when the arm returns with the pendulum, as shown in the second drawing where the pendulum is again central but now moving to the right. The arm follows the pendulum until it strikes the next pin of the escape wheel. Having accompanied the pendulum farther downwards than upwards, it has delivered its gravity impulse and is due for resetting, which is where the neutral arm comes in.

On its continued swing to the right, the pendulum lifts the neutral arm and so releases the escape wheel once again. As the wheel turns, the pin on which the prop is now resting resets the gravity arm to the position shown in the first drawing. When the pendulum swings back to the left, it returns the neutral arm to its banking, and the original position is completely restored. In Big Ben's double three-legged escapement, both of the arms are impulse arms, the lifting of each arm unlocking the train for resetting the other. Apart from that, it is not radically different.

Gravity escapements usually demand more motion from the escape wheel than more conventional escapements, and therefore some extra gearing. An anchor or Graham escapement working with a seconds pendulum requires the escape wheel of thirty teeth to make only a single turn per minute. The crank escapement of Goodrich mentioned in chapter 1 required thirty turns per minute. The single six-legged gravity escapement I have described requires five revolutions per minute and Grimthorpe's double three-legged version requires ten. Because of the extra gearing and the fact that they tend to be noisy, gravity escapements are very rarely found in domestic clocks, but in turret clocks they were a great success. This is because turret clocks need a strong enough drive to cope with large exterior hands exposed to all weathers. If the pendulum is forced to act as a sink for all the surplus power in fine weather, there is too much variation for accurate timekeeping. A gravity escapement isolates the pendulum from the train almost completely, but special arrangements must still be provided to absorb the surplus power from the train.

Big Ben's escape wheel is fitted with a *fly* (like a fan) on the escape wheel arbor, to steady the flight of the train as it resets the gravity arms. This is not all, for the fly itself has inertia and acquires kinetic energy which cannot be comfortably dissipated in an instant when the legs of the escape wheel strike the stops on the gravity arms. The fly is therefore fitted to its arbor with a freewheeling arrangement permitting it to continue turning after the escape wheel has stopped. In small clocks, such as my own W5 described in another chapter, the moderating influence of the fly is purely inertial, owing nothing to air resistance, but it is no less essential and no less effective.

The term *constant force escapement*, sometimes applied to gravity escapements, is one that I have always found highly misleading. It is not intended to imply that the force within each individual impulse is constant. The famous Shortt clock is impulsed by a gravity arm operating on a roller in such a way that a plot of one impulse looks more like a skew volcano than a table top (see Figure 4.2 in the next chapter). There is no possible objection to this, provided the pattern repeats itself exactly. In watches or portable clocks, and in some regulator clocks, constant force escapements use springs instead of gravity arms, as indeed did Froment. Variation in tension as the springs flex and unflex may cause variations within the impulse, but these are precisely repeated from one impulse to the next. As used in horology, constancy of force must be interpreted as constancy of *repetition* of force.

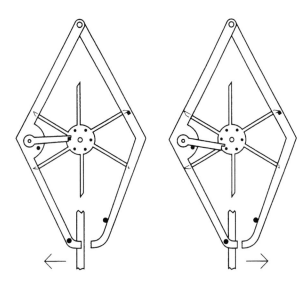

FIGURE 3.9
Arnfield's inertially detached gravity escapement of 1986.

Excellent though they are, gravity escapements cannot really qualify as constant force escapements if the pendulum is called upon to unlock the train. In the escapement already described, the lifting of either arm pulls a stop away from a leg of the escape wheel. The pressure on this stop may only be slight, especially if the legs are long, but it is directly proportional to the torque from the train, so the pendulum is not perfectly isolated. For the ultimate in high precision timekeeping, we would not wish the pendulum to take any part whatsoever in unlocking the clock train. Captain Kater, a name famous in the history of the pendulum, spent the last years of his life trying to perfect an escapement which would avoid this difficulty (he died in 1835), but it is clear from a pathetic report published by his son Edward Kater (1840) that he never succeeded. His method of tackling the problem remained unresolved until James Arnfield (1987) invented his 'inertially detached' gravity escapement for a small skeleton clock.

In Arnfield's clock, the train is released by the gravity arm after it has finished giving impulse and has separated from the pendulum. This was exactly what Kater was trying to do, but the layout of his escapement was less suitable. In form, Arnfield's escapement, Figure 3.9, closely resembles that of the single six-legged gravity escapement, even to the extent of having one impulsing arm (left) and one neutral arm (right), but it works in a different way. The escape wheel does not lock alternately on each side, but only on the neutral arm. The prop which holds up the impulse arm is pivoted. The neutral arm is never touched by the pendulum; it has no pin to engage with the pendulum rod. In the first of the pair of drawings, the pendulum is swinging to the left to pick up the impulse arm. When it does so, the pivoted prop drops to its banking pin, enabling the arm to descend further on the return swing. The impulse ceases when the impulse arm impacts on the neutral arm. The collision holds up the impulse arm sufficiently for the pendulum to continue on its way alone, but gravity is still operating on the impulse arm as it moves the neutral arm aside to unlock the escapement.

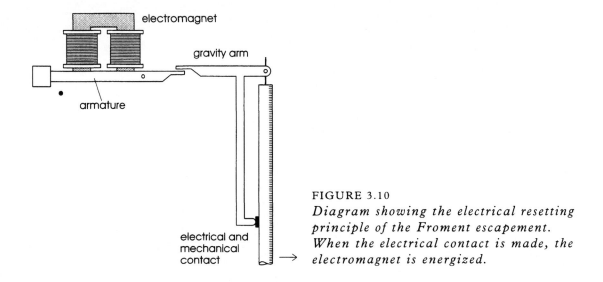

FIGURE 3.10
Diagram showing the electrical resetting principle of the Froment escapement. When the electrical contact is made, the electromagnet is energized.

The unlocking phase is critical to the success of this interesting escapement, as the two arms together must not accelerate so fast that the impulse pin catches up with the pendulum and contacts it a second time in the same swing. It could, in fact, do this almost immediately. The action is so quick that it is impossible to judge what is happening by eye, but by observing the action with a high-speed camera, Arnfield was able to adjust the escapement until the pendulum was completely detached throughout the unlocking. One could only wish that the detachment could be verified in some easier way, perhaps electrically. Similar escapements to this one were made in France in the nineteenth century as Arnfield has pointed out, but it is obvious from their proportions that they were never intended to solve Kater's problem. Their attraction no doubt lay in the provision of deadbeat whole seconds by small clocks with half-second pendulums.

Electricity had solved the unlocking problem at a stroke, though only by abolishing the train altogether and resetting the gravity arm with an electromagnet. This acts as a magnet only when current flows through its coil or 'solenoid'. In Froment's type of escapement, the point of contact between the pendulum rod and the acting end of the gravity arm is also an electrical contact, Figure 3.10. I have not shown the wiring, but while the arm and the rod are in contact, the magnet is energized, otherwise not. (There is no electrical contact where the pivot of the arm appears to coincide with the suspension spring of the pendulum.) When the gravity arm is giving impulse, current flows and the pivoted iron armature is attracted hard up against the electromagnet. At a certain point in the pendulum's swing, the gravity arm is arrested by the tail end of the armature and the impulse is terminated. This breaks the pendulum contact, so the electromagnet ceases to attract and the armature promptly falls away, lifting the gravity arm to a higher position. There it waits to be picked up once again by the pendulum on its return swing; when the pick-up takes place, current flows again and the banking is again lowered. Real versions of this escapement may not look anything like the drawing, which aims only to exhibit the principle.

Switching current on and off is the *sine qua non* of an electrically operated gravity escapement. In 1854, switching necessarily involved making and breaking contacts mechanically, and Grimthorpe would have none of that for Big Ben. He considered electrical contacts too uncertain for use in clockwork, and if my own battery operated doorbell is anything to go by, I have some sympathy with his point of view. However, I decided to take the easy way out and follow in Froment's footsteps in spite of Lord Grimthorpe's crusty objections. After all, my clock would not have quite the same public responsibilities as Big Ben.

CHAPTER FOUR

Echoes of Hope-Jones

Froment's escapement looked simple enough to make if only I could lay hands on an electromagnet. It occurred to me that my electric razor might contain something of the kind and on taking it apart I found just what was needed. It worked perfectly from a low voltage battery with only one drawback – the noise made as the armature snapped hard up against the iron core of the magnet every two seconds. I thought of trying to cushion this with something soft, but any such expedient would make the stroke an uncertain quantity. This would be as bad as a variable force and would invalidate the whole scheme, so the noise would have to be tolerated. Erroneously, I persuaded myself that we would all get used to it, even to the point of missing it when it stopped.

Every constructor experiences a special feeling of excitement when his very own clock shows that it will go and stay going. So it was with 'W1', even without any face or hands. Using Froment's method of impulsing there is no escape wheel to drive the pendulum, and therefore no wheelwork with which to drive the hands. Instead of the escape wheel, there would have to be a *count wheel*, a small ratchet wheel driven a tooth at a time by the swings of the pendulum itself. Gears would then be needed to drive the hands from the count wheel. For this purpose I had few qualms in cannibalizing a cheap dial clock of a type once commonly seen on office walls. Mine had a Smith's EMPIRE movement with what is called a *platform escapement* – a balance and hairspring oscillator mounted on a horizontal platform across the top of the movement. With this removed, along with the mainspring and the heavy end of the gear train, what was left could be used in reverse. All I had to do was to file out a small ratchet wheel and mount it on an existing arbor. A pawl pivoted on the pendulum rod drove this little wheel round, and my clock then told the time. In spite of the pieces manufactured by Smith's, I convinced myself that it was all my own work!

As may well be imagined, the works would best be hidden from view by a dial, for which purpose I begged a sheet of 'half-hard' brass from a workshop in Worcester then busily manufacturing glove-irons for a local factory. My idea was to engrave the dial by hand, and with patience I succeeded, thankful that the brass was no more than half hard! Finally I waxed the grooves and silvered the brass from the recipe for chemical deposition given in Britten's *Handbook* (1938). Having had no education in chemistry that I could remember, it gave me great pleasure converting sticks of silver nitrate purchased from the local store into silver chloride, an operation which had to be carried out in subdued light. Not long afterwards, I had to move house, which was perhaps a blessing in disguise.

There was to be an interesting sequel. Years later I met my old neighbour. Knowing of my interest in clocks, he enquired of my progress and I told him what I had done since the days of W1, forgetting that he had never been told about it in the first place. It transpired that for months his house had suffered from a mysterious clicking noise which repeated at intervals of two seconds throughout the night and seemed to come from the roof, though a thorough search of the loft space had revealed nothing. With some trepidation, I offered him what I described as a possible explanation. Although it had not been fastened to the party wall, W1 had been firmly enough fixed to an exterior wall for the clicks to permeate two four-bedroomed houses. Thankfully, Mr Perry was a scientist and his interest in the technical problem far surpassed any feeling of annoyance at the disturbance. So at least he made it appear, for Mr Perry was not only a scientist but a gentleman.

In a new house, it was unthinkable to fix up W1 again, partly because of the noise, and partly because the driving current flowed continuously for about half of the time, making it extravagant in batteries. A clock like the Synchronome, whose story has been so well told in *Electrical timekeeping* (Hope-Jones 1949), might be the answer and would not be too difficult to make, so I converted W1 into W2. This imitated the principle of the Synchronome clock, but departed from it quite wildly in some of the details. Many amateurs fired by Hope-Jones's enthusiasm have made clocks of this type and in earlier days might have taken advantage of his offer to allow them the use of his firm's workshops in the evenings.

As with Froment, the pendulum of a Synchronome is impulsed by a gravity arm, but only once every half a minute, which I increased to a whole minute so that the peaceful intervals between the clunks of the magnet would be thirty times longer than they had been before. A high Q pendulum can go without an impulse for a minute with the greatest of ease.

Figure 4.1 shows one of Hope-Jones's early designs, patented in 1905. A count wheel of fifteen teeth (thirty in my version) is driven round by a pawl pivoted on a bracket attached to the pendulum rod. The count wheel measures out the half-minute intervals between impulses, during which time the L-shaped gravity arm (pivoted at its corner) is held up on a latch, out of the pendulum's way. In this particular design, the gravity arm is released towards the end of the swing preceding the actual impulse. When the pushing pawl finds itself in a shallower notch of the count wheel, it fails to clear the latch as the pendulum swings to the left. The latch is then pushed aside and the gravity arm released. The arm

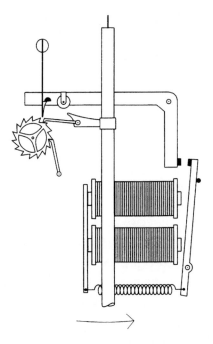

FIGURE 4.1
An early form of Hope-Jones's Synchronome master clock. The two solenoids constitute an electromagnet which is energized when the gravity arm makes contact with the pivoted armature.

does not press directly on the pendulum rod, but is equipped with a little wheel to roll along a curved pallet fixed to the rod. The profile of the pallet is a circle centred at the pendulum's point of suspension, which means that the arm neither rises nor falls as it rolls. This might be called a *rolling rest* analogous to the friction rest in a Graham escapement.

At some stage during the pendulum's return swing to the right, the roller reaches the end of the pallet and the gravity arm falls off. In so doing, it impulses the pendulum before closing the electrical contact on the armature of the electromagnet, causing current to flow and energize the magnet. The magnet then attracts the armature which tosses the gravity arm back on to its sprung latch. Hope-Jones was especially proud of the soundness of this arrangement for various reasons. The electrical contact is made only after the gravity arm and the pendulum have parted company. The mechanical pressure of the contact is quite high because of the leverage exerted by the weight of the arm, enhanced by the force of the resetting action itself. Contact is cleanly broken by the upwards momentum of the gravity arm after the armature has slammed up against the magnet. For me the action never failed, and in my clock – as in all others of that type – Lord Grimthorpe was proved wrong.

The impulse supplied by this design can be central. The actual pattern of the force as the pendulum swings to the right is lopsided, being heavily biased towards the end of the drop, when the roller has greater leverage on the tip of the pallet, Figure 4.2. Hope-Jones does not appear to have understood escapement theory well enough to appreciate that this lopsidedness is of no consequence. Provided a certain condition is satisfied, the time of swing is unaffected by lack of symmetry within the impulse. The mathematical condition

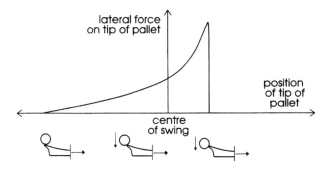

FIGURE 4.2
Graph of the impulse delivered by the falling roller as the pallet swings to the right.

need not be given at this stage; it is sufficient that there is a point within the impulse which can be regarded as its centre, and if that point coincides with the centre of the pendulum's swing, the period is unaffected by the impulse. The escapement error is zero.

This is a matter of some historical interest, for at a later stage Hope-Jones sought the help of W. H. Shortt in removing the lopsidedness of the impulse by extending the pallet and shaping it as an incline for the wheel to roll down. The wheel was not allowed to fall off, but remained on the incline until the electrical contact terminated the impulse. This arrangement was used in all the later Synchronome clocks, but in my opinion it was a step in the wrong direction, for the position of the end of the impulse could now vary with wear on the electrical contacts. Ironically, W. H. Shortt's world-beating observatory clock (described in another chapter) used the lopsided impulse, and yet it was Shortt who, when requested to do so, had calculated the pallet shape needed to give a symmetrical impulse.

The later Synchronome differed from the early design in a more important respect. Instead of pushing the count wheel on swings to the left, it pulled the wheel on swings to the right, enabling the drop of the gravity arm to be delayed until the very swing in which impulse was to be given. In that one swing the arm is released, rolls down its hill, makes contact, and is reset. The rolling rest is eliminated and the pendulum suffers no drag at its extreme of swing, where interference is most damaging to the timekeeping. For my own clock, I ought to have adopted this improvement and combined it with Hope-Jones's original pallet design for free fall of the roller. Foolishly I did neither. Where Hope-Jones had remedied one error only to make another, I made both.

In every Synchronome installation, the electrical circuit for resetting the gravity arm was extended beyond the master clock to include magnets in remote slave dials, stepping their hands on with a click every half a minute. The dial for the master clock was just another slave dial itself. I had no slave dials, so I continued to drive my clock hands from the count wheel as I had done before. The count wheel of W2 thus served a double purpose – to release the gravity arm at intervals of one minute amd also to operate the hands.

That the Synchronome and other similar clocks kept excellent time was almost certainly due to the combination of central impulse and constant force escapement. This may sound a superfluous combination, for if a central impulse eliminates escapement error altogether, why should it be necessary to keep the force of the impulse constant? The answer is that a pendulum is not a naturally isochronous resonator. Wide arcs of swing take slightly longer than narrow ones, making it highly desirable that the amplitude of vibration be kept constant. The pendulum's natural lack of isochronism is usually called *circular error*, though more recently there has been a trend towards calling it *circular deviation*. The older term was thought to sound like a fault, when it is really only a small deviation in the frequency of vibration. However, it becomes a fault if the amplitude of vibration should vary, and I hardly think any change of terminology is necessary. In mechanical clocks, it happens to be quite difficult to ensure a constant amplitude of vibration, so variations in circular error can be serious.

To understand this lack of isochronism intuitively, it is not necessary to work through the mathematics that evaluates it exactly. Think of what would happen if a pendulum were allowed to swing through an extremely large angle, like the church bells used for change ringing. If the pendulum were to swing through half a turn from the centre, it would come to rest upside down and stay there indefinitely! The period of that vibration would be infinite. When the arc of swing is limited to smaller angles, the period is finite. As there is no *abrupt* change of behaviour between these two situations, a pendulum cannot be isochronous, but the lengthening of period with increasing amplitude is actually very small when the amplitude is small. Even for swings of 90° a side, the period is only 18% longer than for small arcs, though that would represent a loss of four hours a day! At an amplitude of 2.00°, the lengthening is no more than 0.0076%, or a loss of 6.58 seconds per day. At 1.99°, the loss is 6.51 seconds per day. The odd six seconds is of no consequence; it is the difference of 0.07 s per day that matters. This may not sound much, but would barely be tolerable in a precision clock. In the precision stakes, therefore, a constant force escapement would seem to be a necessity.

Circular error is often explained as being due to the fact that a pendulum bob moves in an arc of a circle. Whilst this is true, the explanation hardly goes far enough to be helpful. The balance of a chronometer has a circular oscillatory motion, yet it does not suffer from circular error. For the period of a vibration to be independent of its amplitude, there is a well-known condition. The restoring force which tries to return the vibrating element to its central position, must be directly proportional to the distance from the centre. This condition guarantees simple harmonic motion, which is isochronous.

The real source of the trouble with a pendulum is the way the force of gravity is resolved. The geometry of a pendulum divides the downward force of gravity into a component acting along the line of the rod, which is cancelled by the tension in the rod, and a component at right angles to the rod, which acts as a tangential restoring force. This force is not proportional to the angle from the vertical, but to its *sine*. There is a simple approximate formula for the resulting change of period, accurate enough to use in clocks. The maximum angle from the vertical (in radians) is quartered and then squared. The result is the fractional increase of period; for the timekeeping loss in seconds a day, multiply by

86 400. Strangely enough, there is no completely simple way of arriving at this formula. It is a job for the calculus, and not an especially interesting one at that.

The transition from W1 to W2 turned out to be quite simple. I retained the pendulum, gear train and clock-case, and made a new gravity arm with a roller wheel, which was not too difficult an undertaking. As soon as I had scrapped the parts taken from my electric razor, I began to feel more respectable as a horologist! The electromagnet used in W2 consumed much less current because of its highly intermittent action, but it was every bit as noisy. I had fondly imagined that a clank once a minute would be thirty times less disturbing than one every two seconds, but in some ways it is worse as it breaks into a period of silence. In the event, W2 was not to last much longer than W1 had done. Its demise was triggered by a visit from an amateur cellist who wished me to play the piano part of a Brahms sonata. String players have a habit of choosing works with impossible accompaniments, so I hardly noticed the sound of the electromagnet nor even that of the cello as I struggled with the notes. After suffering with the piano on one side and the clock on the other, my sensitive musical friend insisted that the clock be stopped. It was the last time he visited my house and it was the beginning of the end for W2.

'Time by wire' had been the inspiration for clocks like the Synchronome and its contemporaries. One master clock could distribute accurate time to a set of simple slave dials, and a whole office building could be kept synchronized. The simplicity of the system was admirable. To use the principle of such a clock as a stand-alone timekeeper without any slave dials was to miss the whole point of the design. From every angle W2 lacked integrity; part of it was taken from another clock, the escapement was neither a copy nor an original and the workmanship would have done little credit to an apprentice. However, W2 had served its purpose in making me understand how such clocks worked, and now it would have to be scrapped. The pity of it was that it had looked remarkably good from the outside, worked reliably and kept quite good time. For the 107 days of its all-too-brief life, the frequency remained 'stable' (to use the correct word) to one part in a million. In other words the timekeeping was accurate to better than a second a week.

CHAPTER FIVE

Harrison and Congreve

My two electrical clocks had used ratchet wheels to measure out the intervals of one minute between impulses. Could not such a system be used in an all-mechanical clock? In ignorance of horological history I thought this a novel concept, though Congreve had patented such a clock in 1808 and found it a home in Buckingham Palace. I shall return to Congreve later in the chapter. A far more significant figure was John Harrison, now chiefly remembered for his practical proof that timepieces could work accurately enough to solve the problem of finding the longitude at sea. My particular interest was in his so-called grasshopper escapement, which had been invented in about 1722. I wanted to learn all I could from this. If the idea underlying so smooth and silent an escapement could somehow be adapted to infrequent impulsing, I might succeed in making a clock with jumping minutes and no clicks. That was my aim.

The grasshopper escapement must have cast its spell over every mechanically minded person who has ever studied it, yet Rawlings (1993) makes no more than a passing mention of it in his book on the *Science of clocks and watches*. Harrison sought to design maintenance-free precision clocks which would run without oil and remain stable over a long period. The need for oil betrays the presence of friction, that arch-enemy of every watch and clockmaker, especially those tied to the use of brass and steel. The sliding pallets of the old anchor escapement scream out for oil or grease. Harrison's answer to this was to put elbow joints in each of the pallet arms and eliminate the sliding altogether. There remains a little pivot friction, but this is minor by comparison with that of sliding pallets. A hinged door is always easier than a sliding one.

Like so many inventions that seem simple after the event, the idea behind the grasshopper escapement would not have been obvious to ordinary clockmakers of the day. The four simplified drawings in Figure 5.1, show it in its earliest form. The two jointed pallet arms are shown, along with a short portion of the pendulum rod. Where the three meet,

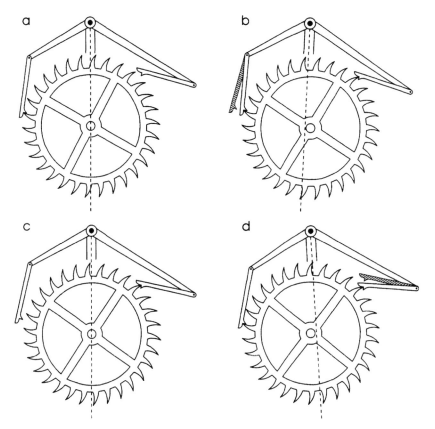

FIGURE 5.1
Harrison's grasshopper escapement in its earliest form.

the angles are fixed; the three parts move as one. Each elbow joint is designed in such a way that the forearm can easily turn on the elbow pivot, but there is a 'natural' angle for the joint, to which it will always return when the forearm is free. How this is contrived need not concern us at the present stage; the important thing is that it happens, and the fact of its happening is the secret of the escapement's action.

In Figure 5.1a, the train is turning the escape wheel clockwise and the pendulum is swinging to the left. It is being helped on its way by the force of the escape wheel tooth acting on the left pallet. The left elbow folds a little during this process. The impulse comes to an end when the right pallet, which is bearing down towards the other side of the wheel, catches a tooth squarely in its notch. This is where the fun starts, for the escape wheel is turning clockwise, and the right pallet will try to reverse it. Something has to give, and it will not be the pendulum. Remember that the pendulum is still swinging to the left and that the upper arms must always move with it. The right elbow recoils the wheel, crumpling a little in the process. As soon as this happens, the left pallet becomes separated from its tooth, Figure 5.1b, and finding itself free, reverts to its natural angle, shown shaded in the drawing. I like to think that this is the *hop* that gave the escapement its

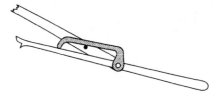

FIGURE 5.2
*Principle of the elbow joint;
the composer is shown shaded.*

nickname, but it has been suggested that the name comes from the resemblance of the pallet arms to the legs of a grasshopper. As the pendulum returns from the left extreme, the wheel resumes its clockwise motion, and is impulsed through the right pallet, Figure 5.1c. The tables are turned when the wheel is recoiled by the arrival of the left pallet, causing the right pallet to hop away, Figure 5.1d. It is a picturesque action. The minimum escaping amplitude for the escapement as proportioned in these drawings is 3° a side.

The elbow joints are fascinating for their simplicity. That of the right pallet, when free, is shown in Figure 5.2. The forearm is counter-balanced to be tail-heavy. At the elbow, there is a third pivoted arm, the *composer*, weighing enough to prevent the arm from rising. The tip of the pallet can be pushed downwards, and when released will return to the composer. If lifted, it will take the composer with it, and when released will again return to the resting position, where the composer is limited by a stop. What a delightfully simple device this is!

My drawings are based on the description by William Laycock (1976) of the escapement which was designed by Harrison for a turret clock in the stables at Brocklesbury Park, Lincolnshire. It seems to have been the only example of its type. A later form of the escapement is that shown in Figure 5.3, where I have again omitted composers and counter-weights for simplicity. This is surely the very pinnacle of escapement elegance, for there is now only a single upper arm moving with the pendulum, and the two pallet arms are pivoted at the same elbow joint. With the composers, there are now five

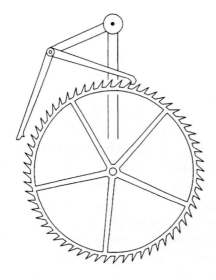

FIGURE 5.3
*The grasshopper in its later
form, with coincident pivots
for the pallet arms. The
composers and counter-weights
are not shown.*

independently moving parts on the one pivot. Note that the left arm is pushed by the escape wheel, and the right arm is pulled. When animated in stages on my computer screen, this drawing also gives an escaping amplitude of three degrees a side, but it is only a drawing to show the principle. The proportions make no pretence at historical accuracy. For a more detailed look at the grasshopper escapement, an article by Peter Hastings (1993) could hardly be bettered.

At the time of writing, a fine new grasshopper escapement of this last type is to be seen working at the British Horological Institute's headquarters at Upton Hall in Nottinghamshire. It is the work of Martin Burgess, and is part of a large sculptural clock commissioned by Barclay's Bank for the City of Norwich, known as the *Gurney clock* to commemorate the foundation of their bank of that name. The proportions of Burgess's escapement are evidently not the same as those shown in my drawing, for its minimum working amplitude is considerably greater than 3°. Harrison is known to have favoured large arcs of swing in order to give the pendulum as much 'dominion' as possible over the clockwork.

Harrison's dominion is not the same thing as a high value of Q, which demands that the energy supplied to the pendulum per period shall be small compared with its total energy of vibration. With a very large arc of swing, the energy of vibration is large, being proportional to the square of the amplitude, but the energy needed to maintain it is correspondingly large. The Q is therefore no greater with a large arc of swing than for a small one, and is in fact slightly reduced because the faster moving bob gives rise to a disproportionate increase in air resistance.

We shall see in chapter 8 that the fundamental virtue of a high Q is the reduction of timekeeping disturbances caused by any slight accidental variations in the effective *phasing* of the impulses. Timekeeping disturbances from this cause are inversely proportional to Q. Variations in the overall *strength* of the impulses are also damaging, because they alter the amplitude of swing. This gives rise to changes of circular error, and – as will be shown in chapter 9 – of escapement error also. Analysis of this problem can be baffling, and it is still a question of some interest as to why Harrison favoured a large amplitude of swing. One explanation is that spurious variations in the force of an impulse can be made smaller in proportion to the total force when that force is relatively large. The combination of a high Q and a large driving force may at first seem contradictory, but it can be achieved in two ways. One is to give large impulses at wide intervals, as in that most accurate of pendulum clocks, the Shortt (chapter 10). Another is to work with a large amplitude of swing and accept the slight loss of Q which that entails. Unfortunately, this aggravates the problem of circular error, which Harrison dealt with by means of rounded cheeks for the suspension spring, as originally proposed by Christiaan Huygens. He also used a spring remontoire to drive the escape wheel and so isolate it from any variations of force from the gearing of the clock train. Both of these features have been reproduced by Martin Burgess in the Gurney clock. The grasshopper does not tick like an ordinary anchor escapement, in which the escape wheel flies out of control for a moment between its escape from one pallet and its check by the next. There is no drop. One advantage claimed for this is that there is no variation in the time at which the impulse starts. Absence of sliding friction is, of course, the grasshopper's prime distinction. In scientific terminology, the pallet friction is static, not kinetic. I know of no technical term to describe the actual hop, in which the frictionless separation of two surfaces brings a new degree of freedom into play. It is only one aspect of the grasshopper, but was just what I was looking for as a way of unlocking an impulse once a minute. From every point of view, it seemed preferable to unfasten a latch or release a detent by separating the surfaces that are under pressure first, rather than by forcing one surface out of the way whilst the other remains in contact with it, which is precisely what Congreve chose to do.

The name of Colonel Sir William Congreve FRS lives on in horology as the designer of a clock controlled by a ball rolling in a zig-zag groove down an inclined plane. Congreve patented this as his own invention in 1808, though a Monsieur Grollier de Servière (1719) relates that his grandfather Nicolas had designed and made just such a clock. Congreve must have been an extraordinary character. At the Royal Laboratory, Woolwich, of which he was eventually to be the Comptroller, he developed the Congreve war rocket, and attended its use in the Napoleonic wars during the period 1807–9. Meanwhile, in 1808 he appeared before King George III in person to be granted the patent (No. 3164) for his rolling ball clock, together with what he called his *extreme detached escapement* for a pendulum. George III was that same monarch who had championed Harrison's cause some 36 years previously, the story of which is well told by Jonathan Betts (1993).

Congreve seemed able to turn his mind to anything, for his inventions included a hydro-pneumatic canal lock and sluice, a process of colour printing which became widely used in Germany, a new form of steam engine, unforgeable banknote paper, a method of killing whales with rockets, improvements to gas meters and a perpetual motion machine based on capillary action! He was a favourite of the Prince Regent, who retained him as his senior equerry on succeeding to the throne as George IV. The contrast with Harrison could hardly be more striking, for both men were full of original ideas, but Congreve had none of the thorny characteristics of a self-made man. He was ebullient and devil-may-care, or so I imagine, and his ideas could be half baked. Both men were awarded the prestigious Copley Medal of the Royal Society.

The notion of *detachment* as a desirable attribute for an escapement had impressed itself on Congreve. A detached escapement is usually taken to mean an escapement which leaves the resonator untouched at its extremes of swing. The detached lever escapement for a watch, further discussed in chapter 10, reduces the frictional loading on the resonator to a very low level, but detachment is seldom applied to pendulums, partly because the pendulum's small angle of swing makes it more difficult to accomplish and partly, no doubt, to the fact that pendulums could keep time well enough with escapements that already existed. None of the conventional escapements for pendulum clocks is detached – other than for a fleeting moment. Having no drop, the grasshopper escapement is never detached for an instant.

Congreve wanted to detach what he called the 'regulating organ' (ie the oscillator) from the escapement, not for some small fraction of time, but for 59 seconds out of 60. This is sound horology and is the line of thought that eventually gave rise to that most accurate of pendulum clocks of the 1920s and 1930s, the Shortt. One of Congreve's proposed regulating organs was the rolling ball, which takes the best part of a minute to traverse a zigzag path down the inclined plane, at the end of which the escapement comes into play for half a second and tips the plane up the other way. Congreve believed that this 'mode of extreme detachment' would be 'of the utmost importance to the final perfection of a true measure of time'. In this he may have been on the right lines, but it was absurd to imagine that it could be done with a rolling ball. It does not seem to have occurred to him that it had no resonance of its own, and that he might just as well have made a water-clock. He seemed to think that because the motion of the ball and a pendulum bob were both under the control of gravity, the two were in some way equivalent. Indeed, his patent specification compares the rolling ball with a pendulum 11 738 feet 4.800 inches in length!

I doubt whether the rolling ball clock ever kept time to better than five minutes a day; an hour a day might be a safer figure to guarantee. If only I could have whispered this word of warning in the ear of certain of today's amateurs who have been beguiled by advertisements for kits of parts for reproducing Congreve's rolling ball clock, much disappointment might have been prevented. In the same patent with the rolling ball fiasco, Congreve describes various other clocks, one of which is a *pendulum* clock receiving impulse once per minute instead of once per swing. He seems to have been unaware that more than fifty years earlier Jean-André Lepaute (1755) had apparently stretched the interval to 15 minutes in a clock whose pendulum was impulsed by its striking mechanism. In Buckingham Palace, Congreve's pendulum clock, which was actually made for him by J. Moxon, must have been under the care of B. L. Vulliamy, who described it as a complicated and troublesome machine (Royer-Collard 1969). Eventually the escapement was scrapped and replaced by something more conventional. The design has been preserved on paper in what must be a treasure house of such failures, the London Patent Office.

The drawing of Congreve's escapement in Figure 5.4 is based on that given in his patent specification. The upper wheel is the count wheel which makes one whole turn per minute, being driven by the clock's seconds pendulum. The lower wheel makes one whole turn per hour and is the escape wheel, equipped with sixty teeth and sixty pins. The minute hand is mounted on its arbor. On taking a closer look, we see that the count wheel is driven by an inverted anchor. The role of the anchor as well as its position are reversed, for instead of driving the pendulum, it is driven by it. The anchor pivot is, of course, in line with the pendulum suspension and its arbor will carry the usual crutch (see Figure 3.4). A third pallet, *P*, on the anchor frame swings in and out of a gap

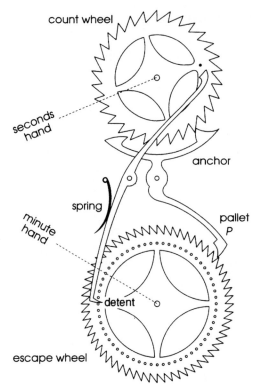

FIGURE 5.4
Congreve's 'extreme detached escapement'.

between two teeth of the locked escape wheel (without touching) for 59 seconds out of 60, so we can forget about this pallet for a while. Of using anchor pallets in reverse, Lord Grimthorpe writes as follows: 'It is perfectly easy to drive a wheel ... by common recoil pallets, provided the wheel is good-natured enough to stand still while the pallets are not moving it.' This hits the nail squarely on the head. To guard against unwanted movement, I would have been tempted to introduce a little deliberate friction, even though this would have reduced the loaded Q of the pendulum.

Separate from the anchor, there is a long arm whose lower end acts as a *detent*, to hold the escape wheel against turning. Once a minute, at an early stage of a leftward swing of the pendulum, a pin in the rim of the count wheel meets the upper end of the long arm, forcing it to turn a trifle clockwise against the force of the leaf spring. The detent is thus pushed out of the way of the escape pins, so releasing the escape wheel. A tooth of the escape wheel then drops on to the pallet P, so impulsing the pendulum through the anchor. Before the pendulum has swung far enough for this tooth to escape, the count wheel's pin must have escaped from the long arm, allowing the detent to obstruct a pin of the escape wheel in time to prevent the disaster of a free run.

In support of his invention, Congreve puts forward some exaggerated claims. He is convinced that the power needed to drive his clock would be little more than one sixtieth of that for any other clock because the escapement is in a state of repose for 59 seconds out of 60. When he goes into the details of his argument, however, it is possible to see where he slips up. The argument rests on two premises, the first of which is that virtually all of a clock's energy is dissipated in overheads, that is, before it ever reaches the pendulum. Most of it is lost, he contends, in the gear train and in the mechanism of the escapement itself. I admire his perspicacity in making this observation, but I think he is too extreme. I have measured the overheads in two of my own clocks and found that only half to three quarters of the total input energy is lost before it reaches the pendulum. His second premise is simply wrong. He thinks that the energy losses are proportional to the time for which the mechanism is active, whereas the losses are far more likely to be proportional to the energy turnover, regardless of how distributed in time. Occasional impulsing saves a factor of 60 in the overall ratio demanded of the gear train, so reducing the number of meshes needed. This undoubtedly reduces the energy loss, but not by a factor of 60.

Fortunately, I did not know about Congreve's extreme detached escapement until I had made several clocks with occasional impulsing. In the world of science, it is considered negligent to solve problems in ignorance of the work of others, but it can sometimes be a blessing in disguise. A sight of Congreve's patent might have tempted me to follow in his footsteps rather than in those of John Harrison. As for the clock by Jean-André Lepaute which was impulsed every fifteen minutes by the striking work, I would have to see it working before I could really believe in it, though the account given in chapter 10 of his *Traité d'horlogerie* seems plain enough.

CHAPTER SIX

Silence for a cellist

The quiet clock I needed for the music room would be my first essay in purely mechanical clockwork, not counting the early Meccano effort, and for obvious reasons I wished to sidestep as many gears as I could. If an escapement could be designed in such a way that the escape wheel would turn in an hour instead of the more usual minute, it should be possible to make an eight-day clock with only a single gear mesh. The need for silence made me abandon the idea of a gravity escapement, and I began to think of Harrison's grasshopper. Perhaps some similar principle could be used for intermittent impulsing.

I would not find it too difficult to make an escape wheel in the form of a pin wheel, Figure 6.1. This would be held at rest by some kind of detent blocking one of the pins while a count wheel measured out the minute. At the minute, the detent could be moved out of the way, preferably not by Congreve's method of simply pushing it aside under pressure, which would be like turning a door handle while leaning against the door, but gently. In the grasshopper, Harrison employed recoil to make a pallet *hop* out of the way, and the same trick might be applied to a detent.

A bracket on the pendulum rod would be fitted with a pivoted arm ending in an impulse pallet in the form of a hook pointing downward between the pins of the escape wheel, Figure 6.2. The arm would be tail-heavy, so that the hook would normally swing clear of the pins. A count wheel would be mounted above the escape wheel to be propelled by a pawl pivoted in the same bracket. The impulse hook would be fitted with a kind of crest, which would be depressed once a minute as it brushed past a pin on the count wheel. The impulse hook would then engage with the wheel and recoil it, causing the detent at the bottom to fall out of engagement. On the return swing (to the right) the escape pin would be pulling at the hook and giving impulse to the pendulum. If something could be added to this design to reset the detent and so arrest the escape wheel at the end of the impulse, the hook would release itself in a Harrisonian hop. So far, that

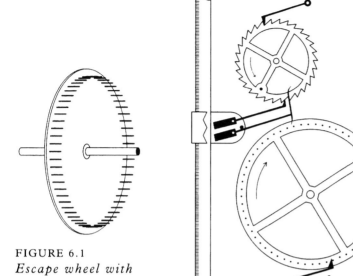

FIGURE 6.1
Escape wheel with sixty pins.

FIGURE 6.2
First thoughts on an escapement for impulsing once a minute.

crucial part was missing. After the first impulse, the pallet would remain hooked to a pin, and the minute hand of the clock would sway to and fro with ever decreasing amplitude, something I have witnessed many times in the course of experimental work.

The solution took a little time to dawn. The detent would have to reset itself automatically. By now I knew in my bones that an intermittent type of grasshopper escapement was within sight, but it still took longer than I thought.

Eventually the detent was made as a bar of rectangular section with two slivers of steel shim let in to its working end, as shown in Figure 6.3. The larger shim is the resetting pallet, and the smaller a safety lip. The resetting pallet must be cut away to leave a passage for the escape pins. The detent is positioned and pivoted much as the one shown in Figure 6.2, and the action is shown in Figure 6.4. For most of the time, the escape wheel is detained by a pin trapped under the lip. When the wheel is recoiled, the pin moves to the right and escapes, causing the detent to fall away. The drop is limited either by a banking pin or by the pallet contacting the next pin of the escape wheel, as shown in the diagram. When the wheel returns from its recoil, this pin lifts the pallet until it reaches the gap, falls through and is trapped in its turn. Here I take the liberty of using the language of relative motion; the pin does not really *fall* through the gap. The gap falls past the pin. The wheel is now locked, and the impulse hook can release itself as previously explained. The escape wheel has now advanced by one pin, and the minute hand has advanced by one division.

As so often happens, a design that is new on paper turns out a little differently in the solid. I jibbed at the thought of making the impulse pallet with a crest, finding it easier to use two separate wires mounted in a common block. The design of the gated detent did

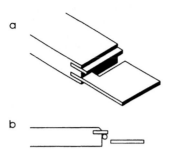

FIGURE 6.3
Detent for escape wheel, showing (a) gate for escape pins, and (b) section through gate.

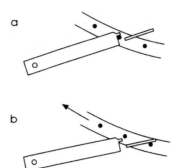

FIGURE 6.4
Action of detent. (a) Escape wheel locked. (b) Detent released by recoil.

not turn out to be critical and needed no alteration. The important point to watch is that the escaped pin can clear the corner of the detent above the lip. As for the lip itself, I believe that with care it might be omitted altogether. When the escape pin reaches the gate, the detent does not have time to fall very far before the pin impacts and stops it from falling away further. The lip is an added security which slightly spoils the principle of the Harrisonian hop.

Although I have called this an *intermittent grasshopper escapement* for want of a better name, it is not friction free. Sliding takes place between the escape pin and the pallet of the detent, but the pressure is only that of the pallet itself, and in principle can be made as light as one pleases. For this reason I made the body of the detent of aluminium. In practice, I have never found more than a light scratching of the steel pallet by the escape pins even after many years running without oil.

The drive line of a weight-driven clock is wound around a barrel which drives the gear train via a ratchet mechanism. The ratchet enables the barrel to be turned back for rewinding, during which the torque exerted on the gear train disappears unless special measures are taken to maintain it. With the intermittent grasshopper some such measure is essential because any momentary absence of torque on the escape wheel causes the detent to fall away, and when the torque is restored, the clock gains a whole minute. To avoid this, I used Harrison's spring-driven maintaining power, which works by means of a spring in series with the power train. The spring, which is kept in a fully compressed state by the force in the train, does nothing but pass on that force while the clock is running normally. However, whilst the barrel is being wound, a fine ratchet with its pawl anchored to the clock frame prevents the maintaining spring from unwinding, enabling it to take over from the barrel as a temporary source of power. As soon as the pressure on the winding key is relaxed, the spring is automatically recharged.

One of the most attractive features of W3 is the motion of the hands. Both the seconds and hour hands exhibit a recoiling action. The seconds hand overshoots the division on the dial and then recoils onto it. The minute hand has a different motion, recoiling before proceeding to the next minute. At the minute, this lively routine caused one kind friend to remark that it made a welcome change from the dead-beat of a Graham escapement.

The purpose of W3 was to avoid any repetition of the episode with my friend the visiting cellist. I could not conceivably have known that the next cellist to visit this room would be the one and only Paul Tortelier, who needed some time for practice before giving a public recital in Great Malvern. He played an unaccompanied Bach suite within a few feet of W3 without a word of complaint. Like fans who cannot bring themselves to wash after being kissed by a pop idol, W3 could never be scrapped after passing muster on such an unforgettable occasion. It was a case of third clock lucky.

When an account of this new escapement was first published in the *Horological Journal*, it was quickly spotted by that master horological artist Martin Burgess, who telephoned me as a complete stranger. 'You know what you have done?' he began. 'You have abolished the train! This is pure Harrison!' I did not know how to answer that, as I had done no such thing in W3. Because I had relied on what I could find in a London junk shop, there were two meshes in the train, not just one as originally intended. Consequently, in order to permit clockwise rewinding, the escape wheel has to turn anti-clockwise. To remedy this, the minute hand is driven through a 1:1 gear mesh, which centralizes the arbor under the count wheel carrying the seconds hand. In addition, there are of course the usual two meshes of motion work to drive the hour hand from the arbor carrying the minute hand. I had therefore finished up with five gear meshes in all! Martin Burgess's compliment tempted me to think of trying all over again.

CHAPTER SEVEN

Going without gears

In response to a few turns of the winding key once a week, the escape wheel of an ordinary mechanical clock has to make at least ten thousand turns. That ratio, about a thousand to one, could be achieved with a single gear mesh by using a wheel 20 feet in diameter engaging with a quarter inch pinion, though none but an eccentric designer of sculptural skeleton clocks would consider such a possibility. It never ceases to surprise me that the same ratio can be achieved in three meshes using gears only 2.5 inches in diameter with three of the same quarter inch pinions. It is of course the multiplicative property of gear ratios that makes a cascade of gears so effective, and there is the added advantage that intermediate arbors can be made to turn at the correct speeds for driving hour and minute hands. In a traditional regulator clock this is just what happens; the second, minute and hour hands are all on arbors of the power train. In more ordinary clocks, the minute hand is on an arbor of the train, whilst the hour hand is driven from a branch line of two meshes called *motion work*, which doubles back on itself to make the hands concentric, Figure 7.1. Motion work is just what it says: it conveys motion without power. As a result it is apt to be sloppy, because there is no tension to take up any play in the gears.

The plan of the traditional regulator clock is so neat and economical that it would seem obtuse to consider anything else, and yet I was still determined to attempt a clock with no meshing gears at all. This would necessarily involve driving the escape arbor directly. My aim would have seemed less eccentric had I been aware that Pierre Le Roy and Jean-André Lepaute had presented a clock of a similar kind to King Louis XV in 1751! Describing it in his *Traité d'horlogerie*, Lepaute (1755) says that the idea was suggested to him by Le Roy but that it was far from being in the state of perfection to which he, Lepaute, brought it after

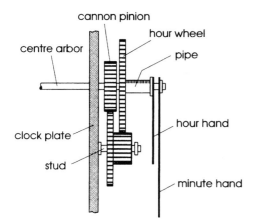

FIGURE 7.1
Traditional motion work. The centre arbor, which drives the minute hand, carries a pinion to drive the gearing for the hour hand, mounted on a pipe.

repeated trials and infinite pain.* A summary of Lepaute's account of his work is given in English by Paul Chamberlain (1941) in that most useful and readable book *It's about time*. A somewhat later example of this clock, bearing the inscription 'Inventée et Executée par Le Roy fils' on its glass dial, can be seen by appointment at the British Museum.

Le Roy's clock is driven by an endless chain hanging in two loops, one for the driving weight and one for a small counter-weight, Figure 7.2, both with pulleys. The chain is draped over a small sprocket wheel to drive the escapement, and a larger sprocket wheel connected to nothing but a ratchet. The small wheels on the left and right are freely turning rollers to position the chain in a pleasing two-dimensional pattern. The use of an endless chain is usually credited to Huygens, its great virtue being that the clock can be wound up by pulling the chain without releasing the driving tension, so no special maintaining power is needed.

The links of Le Roy's chain have a pitch of just 2 mm, and the driving sprocket has six spikes around its circumference to catch alternate links. The effective diameter of the sprocket wheel is therefore an extraordinary 7.6 millimetres. It has to be small because the rate of turning is governed by the escapement with no gear train to reduce it. Without the ingenuity Le Roy exercised, he would have found it impossible to avoid a quite excessive fall of the driving weight to keep the clock going for a week. In fact, it goes for about eight days with a fall of 1.5 metres. (The fall shown in the drawing would correspond to only 0.36 metres.) To take maximum advantage of the fall available in an ordinary long case, Le Roy embedded the pulley of the driving weight within the weight itself.

---

* 'Au commencement de 1751 le fils aîné de M. Julien le Roy me proposa de faire une Pendule dans laquelle il n'y auroit qu'un rateau qui seroit retenu par des échellons fixés sur le balancier; la roüe des minutes devoit pousser le rateau pendant une minute, & le rateau devoit revenir par son propre poids, ou par le moyen d'un contrepoids, pendant la minute suivante; le pendule placé sur le côté de la roüe devoit recevoir toutes les 2 minutes, un coup d'une dent de la roüe des minutes pour lui restituer le mouvement; cette idée étoit encore fort éloignée de l'état de perfection où je la portai bien-tôt, après des tentatives réitérées & des peines infinies, pour la mettre en l'état où nous la présentâmes au Roi dans le mois de Mai 1751' (Lepaute, loc. cit.).

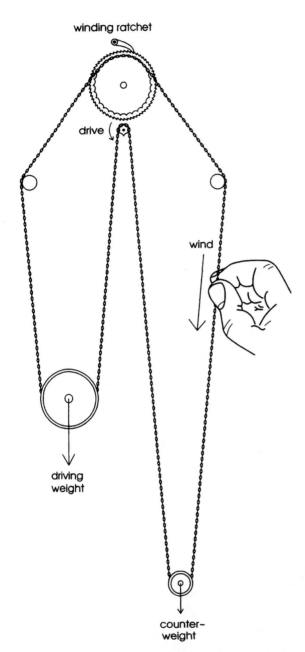

FIGURE 7.2
Huygens' endless driving chain as arranged in Le Roy's 'one-wheel' clock. The chain is much shortened in this drawing.

The key question is how the escape wheel can be made to turn slowly and yet drive the pendulum. Lepaute's treatise includes a design using count wheels to measure out wide intervals between impulses more than fifty years before Congreve, but Le Roy's solution to the escapement problem is quite different. His *tour de force* was the use of two escapements in series, an idea that would never have occurred to me in a month of Sundays.

Le Roy's 'one-wheel' clock. The seconds are indicated by a to-and-fro pointer under the dial, connected to the rake of the secondary escapement which can be seen between the numerals 55 and 60 on the dial.

Instead of driving a crutch, the primary escapement with its duplex wheel of ninety teeth drives a 'rateau' (rake) from side to side in very slow time, each swing taking thirty seconds. In Figure 7.3, I have masked out the details of the primary escapement, which would be too small to study. The secondary escapement involves the rake, which carries fifteen pins in this particular clock, and these weave their way between a symmetrical trio of pallets on an arm jutting from the top of the pendulum. Once more, I find myself using the language of relative motion, for it is really the pallets that do the weaving. As the arm nods up and down, the pins give impulse to the pendulum on each swing, no matter which way the rake is being impelled. In the enlarged detail, Figure 7.4, I have used artistic licence and made the rake transparent to show the pins and pallets.

The primary escapement has been described by other writers as a form of duplex escapement, which is thoroughly misleading to those familiar with watch escapements. Although I have not had the opportunity to take it apart and see the relevant pieces that

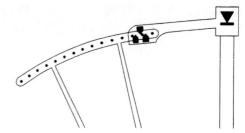

FIGURE 7.4
Detail of secondary escapement.

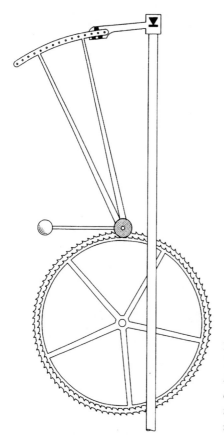

FIGURE 7.3
Two-tier escapement of Le Roy's one-wheel clock; the primary escapement is operated by the duplex escape wheel and the secondary by the to-and-fro motion of an arc of pins. The knife-edge suspension is here represented diagrammatically.

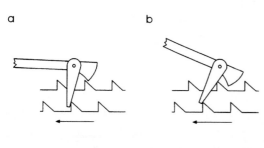

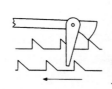

FIGURE 7.5
Conjectured form of Le Roy's primary escapement. The broken arm carries a counterpoise. (a) Inner scape is lifting the counterpoise. (b) End of lift. (c) Outer scape assists fall of counterpoise. (d) End of fall.

are hidden from view in the clock, computer simulation – coupled with an engraving from Lepaute's book – convinces me that the series of diagrams, Figure 7.5, is not far from the truth. The pallet arbor carries a curved pallet shaped like the blade of an axe and a straight pallet in a different plane. Both are rigidly fastened to the arbor, along with an arm having a counter-weight on the end of it, omitted in these diagrams. The radii supporting the rake are of course driven from this pallet arbor, and the action is as follows.

The escape wheel, with two sets of ninety teeth fixed together, is trying to turn anti-clockwise. In Figure 7.5a, a tooth of the inner set is pushing the straight pallet to the left, which drives the rake to the right and lifts the counter-weight on the end of the broken-ended arm. In Figure 7.5b, the tooth is about to escape, after which a tooth of the outer set will drop on to the curved pallet. In Figure 7.5c, the slight inward spiral of the curved pallet is assisting the slow fall of the counter-weight which is driving the rake back to the left. At Figure 7.5d, that tooth is about to escape, after which the straight pallet comes back into operation.

In the duplex escapement of a watch, which is more subtle, the part I am calling the curved pallet gives no impulse at all. It is actually a cylindrical roller providing a friction rest for the supplementary arc and for the whole of the return swing. In Le Roy's clock there is no supplementary arc, and the curved pallet could be shaped to spiral inwards. By comparison with the direct thrust on the straight pallet, however, this is a poor arrangement with little leverage and much friction, as Lepaute complains in his treatise. In proportioning the pallets for my drawings, it seemed to me that little if any of the escape wheel's travel should be wasted on the curved pallet, as the counter-weight can provide as much force as necessary, and this will be recharged on the next stroke by the other more efficient pallet.

Le Roy could not use the escape wheel, which takes ninety minutes to turn, to drive the minute hand directly. Instead, he uses the escape arbor as the intermediate arbor of his motion work. In other words, there is one gear mesh of ratio 3/2 from the escape wheel to the minute hand and another of ratio 1/8 to the hour hand. Full of admiration as I am for the ingenuity of Le Roy's two-tier escapement, I am not attracted to it from a practical point of view. It must surely be difficult to adjust, and the power consumption works out at 60 microwatts, as against 5 microwatts for the gearless clock I am about to describe.

To drive an escapement without even a single gear mesh is one thing, but quite another is to find a means of winding the clock up. I had decided to use a revised version of the quiet escapement described in the previous chapter, in which the escape wheel makes one turn per hour. There are 168 hours in a week, which effectively rules out the use of a key or even a crank handle for rewinding. (One kind friend who admitted that he was no purist suggested the use of an electric motor …) Although I did not know of Le Roy's clock at this time, I had considered the use of a Huygens endless line but had to dismiss the idea as too difficult with the fine line that would be necessary. There would have to be some kind of friction drive from a smooth line with two ends, wound around the escape wheel arbor. The

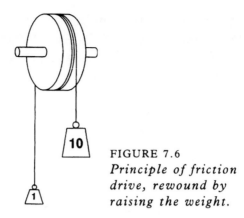

FIGURE 7.6
Principle of friction drive, rewound by raising the weight.

principle of the bollard has long fascinated me, with its polished surface gripping a slippery rope so tightly that a heavy boat can be constrained with no more than a hand hold.

The principle seemed clear enough in theory, though not in practice. The idea was to hang a heavy weight on one end and a lighter weight on the other, as in Figure 7.6. The line would turn the arbor without slipping, and would be rewound by raising the driving weight, not by pulling the counter-weight down, which would be like trying to pull a boat in while its painter was wound round a fixed bollard. After much inconclusive fiddling with anglers' line and rods of brass, I had to abandon most of the bollard. On lifting the weight, the coils would cross over one another and jam, but a single turn did not provide enough friction until I thought of using a V-pulley.

The pulley idea was successful, provided that the line was wrapped round almost a full turn. To be certain, I had carried out a series of experiments with different angles of wrap. For the benefit of other experimenters who may be interested, the line was 0.2 mm nylon monofilament, and the pulley of polished steel had a diameter of 4 mm at the inner point of the V, whose angle was 45°. With a quarter turn, I found that the counterweight had to be at least four fifths as heavy as the driving weight; with half a turn, the figure was two thirds, and with three quarters about a half. For safety, I chose to work with seven eighths of a turn and a counterweight 60% of the driving weight, but it is still not a good return for the mass of metal required. If the pull needed by the clock were 1 kg, the tension on the driving side of the pulley would have to be 2.5 kg, with 1.5 kg on the other side. Without pulleys, therefore, a total of 4 kg would be needed to provide a working pull of 1 kg. With a pair of pulleys, one for each weight, the mass would be 8 kg.

At this stage, I had very little idea what tension would be needed, but in the event it turned out to be a modest 120 grams. The tension on the driving side of the pulley is nominally 300 g and that on the driven side 180 g. The driving weight weighs 600 g, being suspended by a pulley, but for aesthetic reasons the counter-weight has no pulley and therefore weighs 180 g. An outline of the drive is shown in Figure 7.7, with its 10-inch dial (dashed circle) and 0.8 Hz pendulum, i.e. 48 vibrations or 96 swings per minute. The line can be seen almost encircling the escape wheel arbor, which is in the centre of the clock

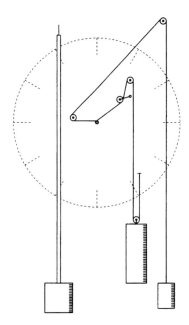

FIGURE 7.7
Outline design for gearless clock, showing the drive line wrapped around the central arbor of the clock. A jockey pulley is used to actuate maintaining power during winding.

and carries the minute hand. As drawn, the line is adjusted for thirty-hour going; for eight-day going, it would be longer and a pulley would be needed for the counter-weight.

Some means had to be found to maintain the torque on the escape wheel during winding. As the weight is lifted by hand, the tension on the drive line is much reduced, and this fact can be exploited. The drawing shows a 'jockey pulley' on a short pivoted arm, as though taking up slack. During winding, the slack is greater, causing the arm supporting the pulley to turn a little. It is a simple matter to arrange for this motion to lower a pressure pad on to one of the escape wheel's pins to maintain the required torque, Figure 7.8. It was too much to hope that this force would be exactly right. Unlike most maintainers that seem too weak, this one was far too strong. The excessive torque on the escape wheel could even stop the clock, as the pendulum had not sufficient inertia to recoil the wheel and unlock the detent. This explains the use of a secondary arm with a weaker spring resting on the heavier

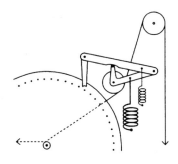

FIGURE 7.8
Detail of jockey pulley controlling a pallet hanging over the escape wheel.

FIGURE 7.9
The author's gearless clock with dials removed.

arm that carries the pulley. Although the maintaining power was primarily intended to keep the detent engaged during winding, it can in fact deliver one impulse during winding.

In the revised form of the escapement, Figures 7.9 and 7.10, the count wheel turns anti-clockwise, making its arbor unsuitable for a seconds hand. The usual answer to such a predicament is to have a fixed pointer pointing to numerals on the wheel itself, which for some reason makes it appear as though the wheel is turning in the correct sense. I was able to combine this feature with an aperture in the dial to show off the whole of the escapement mechanism without compromising the legibility of the hours and minutes. The action of the escapement should be almost self-explanatory. One tooth of the count wheel is cut more deeply than the others, allowing the pawl to catch the end of a lever L which deflects the impulse hook. Parts shown on the drawing as a single thick line are made of stainless steel wire. The detent can be placed anywhere around the periphery of the escape wheel, and was placed at the top of the wheel where it needs to be made tail-heavy, but where it can be seen through the aperture in the dial.

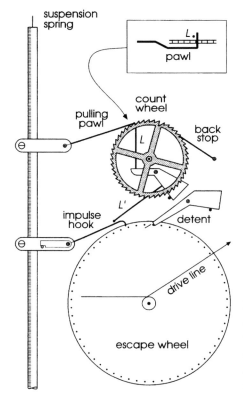

FIGURE 7.10
Escapement of gearless clock. The pawl on the upper bracket of the pendulum rod catches the lever L once a minute. The other end of the lever, L' causes the impulse hook to engage with a pin of the escape wheel.

Craftsmen clockmakers may frown at the use of bent wire, though two eminent makers have been gracious enough to complement me on it. In traditional horology, rigid parts are carved from the solid, and can be admired as works of sculpture. Bent stainless steel wire is easier and has much to commend it; weight for weight its resistance to distortion is far superior to that of a piece made of solid brass or soft iron. I make my count wheels from Perspex® acrylic sheet, partly for its lightness but mainly because at light pressures it resists wear from the stainless steel pawls. In over twenty years running of the earlier clock W3, I have detected no sign of wear on the count wheel, though very light lubrication seems to be needed to prevent sticking.

Gears being barred by my self-imposed rules, the hour hand was driven by a cam arrangement on the central arbor. The drawing in Figure 7.11 is not based on the actual design, but shows the idea. An elliptical cam C is mounted on the central arbor along with the escape wheel and the minute hand, and a ratchet wheel is mounted on a separate arbor for a separate (non-concentric) hour hand. A cam follower pivoted at P carries a dangling arm which acts as a pawl to drive the ratchet wheel, whose backstop is pivoted at B. In twelve hours, the follower rocks up and down twenty-four times, and the ratchet has twenty-four teeth.

The author's gearless clock, whose escapement can be seen through an aperture in the mahogany face. The small weight in the top left corner is cantilevered from the plain steel pendulum rod by a bimetallic strip for temperature compensation.

With this simple arrangement, the hour hand would pause for quarter of an hour in every half – which would hardly matter – but such pauses are easily eliminated. Imagine the cam to be halved in size, so that the pawl traverses only half a tooth at a time. The backstop is made to rock up and down from a separate cam, so that it works as a second pawl. The two cams are mounted at right angles on the same arbor, and the pawls phased so that as soon as one push stops, the other one starts, making the motion of the hour hand remarkably smooth. Using cranks in place of cams, a similar push–pull system is used by Peter Brain to convey circular motion to distant dial work in his reproduction of Lord Kelvin's free pendulum clock (chapter 17).

Were I ever to make another clock of this type, I would probably introduce one gear mesh in the main drive, and I would have another think about the motion work, thanks to an unexpected visit from Charles Brandram Jones, who had been reading an article about my gearless clock. In the course of a most stimulating discussion, he drew my attention to the astonishingly simple motion work invented in the United States by Aaron Dodd

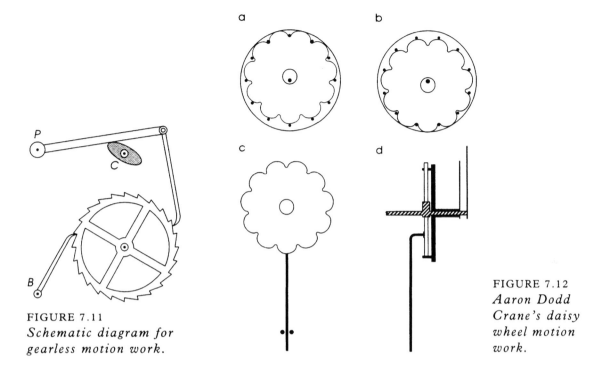

FIGURE 7.11
Schematic diagram for gearless motion work.

FIGURE 7.12
Aaron Dodd Crane's daisy wheel motion work.

Crane, whose little known and highly original clockwork has been admirably described by Frederick Shelley (1987).

Crane's 'daisy wheel' motion work uses only the central arbor of the clock (which always carries the minute hand). On this arbor, there is the usual surrounding pipe for the hour hand but no other arbor. To any watch or clockmaker, this may sound magical. So indeed it is, and it is not to be grasped in a hurry! The first drawing, in Figure 7.12, is a purely mathematical sketch showing a hubless disc with eleven petals – the daisy wheel – resting on a circular disc with twelve pins, which is the hour wheel. If the daisy wheel were to be held fixed, the hour wheel could be rolled around it, each pin in succession slipping down into the gap between two petals, advancing the pin wheel gradually in relation to the daisy wheel. After one full roll around, the drawing would look exactly the same, but the pin which was originally at the top would occupy the one o'clock position. The second drawing, Figure 7.12b, shows the half-way stage. The principle is that of epicyclic gearing with teeth of an unorthodox shape. The pin wheel is in fact a kind of lantern pinion engaging with an internal gear wheel.

To make this system work, the daisy wheel is mounted loosely on an eccentric collar rigidly fixed on the central arbor of the clock. When the central arbor turns, the collar turns and causes the daisy wheel to roll inside the pin wheel. As both wheels are floating on their axes, there has of course to be some restraint, so the daisy wheel is prevented from rotating on its own axis by means of a stalk shown in Figure 7.12c. This stalk is fastened rigidly to the wheel and constrained loosely between a couple of pins, leaving enough freedom for the rolling action without permitting the wheel to rotate on its own axis. A sideways view is shown at Figure 7.12d, where the pin wheel on its hour pipe is shown in black, the arbor with its eccentric collar is hatched, and the daisy wheel is unshaded. Compare this, for simplicity and elegance, with the traditional motion work in Figure 7.1.

With an hour wheel of twelve pins, the explanation is reasonably straightforward, but Crane used a spider construction with fewer pins. It should be clear from Figure 7.12a that the removal of every other pin would hardly upset things, and it seems that four or even three pins were quite enough. Faced with the former of these possibilities, and with no further explanation, antiquarians visiting the turret clock in the Phenix Baptist Church at West Warwick, Rhode Island, might be forgiven for finding the principle elusive. In Frederick Shelley's words, 'it is one of those fantastic innovations that must be seen to be believed'. Whether or not it counts as a gear is not a matter of any great consequence.

CHAPTER EIGHT

Disturbed harmonic motion

At the age of twenty-five, George Airy (1827) wrote a paper on escapement theory which has stood the test of time and become a classic work. Two months later he was appointed Lucasian Professor of Mathematics at Cambridge University. In 1835 he became Astronomer Royal and in 1871 President of the Royal Society. Today, Airy's paper would be regarded by mathematical physicists as quite elementary, but few things in science are elementary when being done for the first time.

Regrettably, scientific papers written more than a century ago are seldom read in the original, because information of lasting value has always worked its way into the fabric of scientific knowledge with or without any reference to the originator's name. Horologists do not lightly discard famous names from the past, and one or two of the truths in Airy's paper seem to have become known as *Airy's Laws*. I am by no means sure that I know which these are. Airy considered a single period of vibration in which a small extraneous force is introduced, over and above the restoring force. It might be a frictional force or it might be the force from a clock's escapement. In practice, of course, forces of both types are present simultaneously, but Airy's force was arbitrary, and he derived two fundamental formulae expressed in the language of the integral calculus to show what effect it would have on the amplitude and period of vibration. Then, as examples in the application of the formulae, he worked his way effortlessly through nine horological problems. Over the years, one or two of these examples must have been quoted sufficiently frequently to be accorded legal status, but in reality Airy invented no new laws. He simply applied those of Newton.

In his example 4, Airy shows how any frictional force whose instantaneous strength depends only on the velocity of the pendulum, will, in the course of one period, diminish the arc of swing without altering its period. Turning to escapements, in his example 8 he considers a force which is 'equal at equal distances from the lowest point on both sides',

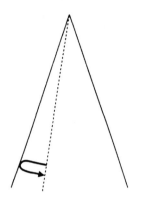

FIGURE 8.1
One of Airy's examples: a force acting towards the centre, equally during the outward and return swings of a pendulum, supplies no energy but shortens the period.

and finds that 'the action before reaching the lowest point tends to diminish the time of oscillation, and after it to increase the time, and that on the whole the time of oscillation is not altered'. This is important. A force directed towards the centre shortens the vibration, a force directed away from it lengthens it. A force disposed symmetrically about the centre introduces no change. Airy's example 9 deals with 'a force which is equal at equal distances as the pendulum ascends and descends to the same place'. By this, he means a force acting through the extreme of swing, Figure 8.1, such as that which precedes the main impulse delivered by a recoil escapement. He concludes that 'a force of this kind does not alter the arc of vibration, but tends during the whole of its action to diminish the time'. In other words, it supplies no energy but causes the clock to gain.

Notions such as these would scarcely have been news to Harrison a century earlier. Indeed, Harrison modified the distribution of force during the recoil of his grasshopper escapement to reduce the 'diminution of time' and made important modifications to the design of the verge escapement for use in his prizewinning timepiece H4 to the same end. He must have known that escapement error varies with amplitude of swing, and took appropriate steps to keep the amplitude constant. It is doubtful whether Airy could have taught Harrison anything at all on this subject, but Harrison could not teach others. It took an Airy to write down the theory of escapement error in the exact language of science, so that any persons with a knowledge of calculus could work it out for themselves.

Airy's theory is not exact. It is an approximation based on the assumption that Q is very large, an assumption he had perforce to express differently because the parameter Q had not yet been defined. Approximations are always apt to become stumbling blocks for those untrained in applied mathematics, which is full of examples where some small effect is evaluated, neglecting what are called second order terms.

A simple example is to be found in the rule for lengthening a pendulum to make a clock lose by some fraction, such as 86.4 seconds a day, which is one part in 1000. The simple rule is to lengthen the pendulum by two parts in 1000. This rule follows from the fact that the period is proportional to the square root of the length, or in other words the length is proportional to the square of the period. However, the square of 1.001 is actually 1.002 001, so we should really lengthen the pendulum not by two parts in 1000, but by

2.001 parts. By ignoring that thousandth of a part, we are only making a second order error, a small error in an error that was already small. Much the same thing happens in escapement theory – though in a more interesting way – and the resulting second order error is inversely proportional to the square of Q. For a pendulum with a Q of 10 000, therefore, the final error in Airy's theory is less than one part in a hundred million. The approximation made in escapement theory involves not just a single number but a complete wave shape. In Airy's approach, the shape of any individual vibration is assumed to be unaffected by the very disturbance which alters the period and the amplitude. It is a move which laymen invariably find hard to swallow, though the error introduced is only of the second order.

In 1988 the horological fraternity in Britain started to take a renewed and sometimes querulous interest in Airy's work, as a consequence of which I set about the task of presenting the theory in a more modern guise (Woodward 1989). I confined attention to the pendulum in a steady state of oscillation, in which the amplitude of swing has settled to a steady value with the energy gains and losses balanced, as happens in every running clock. By assuming such a steady state from the outset, the generality of Airy's theory is sacrificed, but some conceptual difficulties are removed. A further simplification throughout the present chapter is to confine attention to sinusoidal forces, as this postpones the point at which second order terms have to be ignored, so turning what many people consider to be a difficult subject into a relatively easy one.

I make no pretence at what mathematicians call 'rigour' in the following presentation of escapement theory, nor do I pretend that my approach is in any way superior to that of Airy's original paper. It is intended more as a description of what escapement theory is about, and what it has to tell us. The conclusion is summarized in a single rule, which will appear in equation (10). Readers who dislike equations can take comfort in their simplicity. There is no calculus. To follow this account, one has to know the meanings of a sine and a cosine, and to remember (or accept) the rule which tells us the cosine of a difference of two angles. I doubt whether escapement theory can be reduced to terms much simpler than this.

Sinusoidal motion is a projection of circular motion. When we watch the pedals of a bicycle from behind, they appear to go up and down, and the motion appears sinusoidal. With a slight effort of the imagination, the brain can be persuaded to forget the circular motion altogether, real though it is. A pendulum bob challenges the imagination in the opposite way. The almost linear motion we see as the bob of a grandfather clock swings from side to side can be imagined, admittedly with some difficulty, as a projection of circular motion. The fictitious circle is in a horizontal plane, level with the bob, and has nothing to do with the pendulum's own circular arc of swing, which is an unfortunate complication in this exercise. To avoid it, forget about the pendulum rod for the time being and think of the bob as a lump of metal standing on an ice rink, constrained left and right by a pair of springs. The motion would be similar to that of a suspended bob but the swinging would then be in a straight line, as seen from the edge of the rink. The drawing

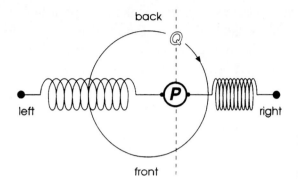

FIGURE 8.2
The springs cause P to oscillate; the imaginary point Q traces out a circle.

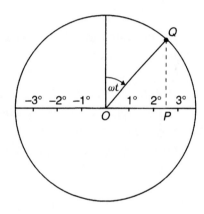

FIGURE 8.3
Phase diagram for pendulum swinging with amplitude 3.75°.

in Figure 8.2 is a view looking downwards from above, showing the real bob P and the fictitious circulating bob Q.

The purpose of this construction is to tell us the position of the bob at any instant. An experienced clockmaker once asked me that very question. Knowing that his seconds pendulum took half a second to go from centre to extreme and half a second to return, he was puzzled that he did not know how long it took to reach an intermediate point he happened to be interested in. With a little trigonometry, the 'phase circle' answers this question readily. Taking (say) 3.75° as the pendulum's amplitude of swing, we mark out a horizontal scale from −3.75° to 3.75° and draw a circle with this as diameter, Figure 8.3. A point on the phase circle is imagined to be orbiting clockwise at a constant angular velocity ω radians per second, taking a time T to make one revolution (i.e. 2 seconds for a pendulum beating seconds). The symbol ω is simply short for $2\pi/T$. If we measure time from the moment the pendulum passes through centre swinging to the right, the formula for the position of P is

$$P = p\sin\omega t. \tag{1}$$

where p is the amplitude, here taken to be 3.75°, and ωt is the *phase angle*. The time taken for the bob to reach an arc angle of 2.5° (say) to the right of centre, can be found by solv-

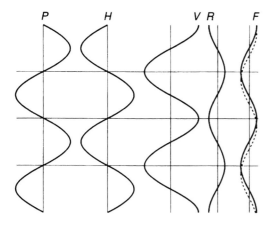

FIGURE 8.4
Sinusoidal variations of position P, homing force H, velocity V, resistance R, and applied force F (broken curve mistimed).

ing equation (1) for t. Thus, putting $P = 2.5$ and $p = 3.75$, we find $\sin \omega t = 2.5/3.75$, from which the phase angle is found to be arcsin (2/3), or about 42°. A seconds pendulum takes one second to make half a revolution on the phase diagram, giving as the result $t = 42/180 = 0.23$ seconds approximately.

My approach to escapement theory involves plotting – or imagining – several different sinusoidal curves, the first of which is the graph of equation (1), showing the lateral position of the bob as it varies with time plotted downwards on the paper, Figure 8.4. Two periods are shown, the horizontal lines marking out half periods. Other sinusoids include the *homing force H*, otherwise known as the restoring force, which can be represented with amplitude h as

$$H = -h \sin \omega t. \qquad (2)$$

This has a minus because it is in anti-phase with position. When the pendulum is on the right, the force acts towards the left, and vice versa. The velocity of the pendulum (V in the figure) is also sinusoidal, but it goes as $\cos \omega t$. When the position is central, the velocity is at a peak value, and when the velocity is zero, the position is at one of its extremes. Position and velocity are said to be in quadrature, because the phases of the two curves differ by a right angle. The phase of the velocity V is in fact 90° *ahead* of the phase of the position P.

There are two other forces to take into account, firstly the *resistive force R*, which can be represented with amplitude r as

$$R = -r \cos \omega t. \qquad (3)$$

The minus is needed here because resistance is in anti-phase with velocity. When the pendulum is moving from left to right, the resistive force opposes it by acting from right to left. In practice the resistive force would be so small in relation to the homing force as to be invisible when drawn on the same scale. It is in fact Q times smaller, i.e.

$$h = Qr. \qquad (4)$$

This is the same Q as has been discussed in chapter 2. The scale of R in Figure 8.4 has perforce been grossly exaggerated.

Finally there is the driving force to consider, for without any drive the clock would stop. Ideally, the drive would be equal and opposite to the resistive force, not just on average but at every instant, so we would have

$$F = f\cos\omega t, \qquad (5)$$

and f would be equal to the r in equation (3). Thus $F + R$ would be identically zero and there would be no interference whatsoever with the timekeeping. However, if the driving force happens not to be phased in this ideal manner, but slightly delayed as indicated by the broken curve in Figure 8.4, some interference does take place, and that is what we must try to evaluate.

Let us therefore introduce a phase lag λ into the timing of the drive, whilst leaving its amplitude f unchanged. The drive may now be represented by the formula

$$\begin{aligned} F' &= f.\cos(\omega t - \lambda) \\ &= f.\cos\lambda\cos\omega t + f.\sin\lambda\sin\omega t. \end{aligned} \qquad (6)$$

The applied force has been split into two parts, the first varying as $\cos\omega t$ with amplitude $f\cos\lambda$ and the second varying as $\sin\omega t$ with amplitude $f\sin\lambda$. The first of these has the 'correct' phase, i.e. the same phase as F had, and this is the energy-giving part of the drive. Its strength is slightly reduced by the factor $\cos\lambda$, which is less than 1. However, provided the clock does not stop in consequence, the pendulum's amplitude will settle to a new and smaller value at which the balance

$$f\cos\lambda = r \qquad (7)$$

holds good. So much for the effect on the amplitude. The second term in equation (6) is the fly in the ointment. Being proportional to $\sin\omega t$, it is in anti-phase with the homing force given in equation (2) and will subtract from it. This does nothing to the energy balance. A pure homing force never does. In the language of electronics, it is a *reactive* force; what it does do is to alter the period of swing. The reduced homing force appears to the pendulum like a reduction in the force of gravity, causing the clock to lose, and we must now see by how much.

The extra term reduces the amplitude of the homing force to $h - f\sin\lambda$, which is a fractional reduction of

$$\frac{f \sin \lambda}{h}. \tag{8}$$

By equation (4), this is equal to

$$\frac{f \sin \lambda}{rQ}. \tag{9}$$

We now substitute for f from equation (7) and, reminding ourselves that tan = sin/cos, find that the fractional reduction of the homing force is simply $(\tan \lambda)/Q$. As frequency of vibration is proportional to the square root of the homing force, the fractional loss of time is (for small fractions) half the fractional reduction of the force, finally giving the result

$$\text{fractional loss of time} = \frac{\tan \lambda}{2Q}. \tag{10}$$

This is the principal result of Airy's escapement theory when applied to a pendulum which has settled down to a steady amplitude of swing, and is the goal we have been striving to reach. It is usually called the *escapement error*, in which context I like to refer to it as the *tangent rule*.

Looking at this rule, we can draw some useful conclusions immediately. If the applied force is in phase with the velocity of the pendulum, the period of vibration is unaffected, because λ and $\tan \lambda$ are both zero. This we knew already. If the force is applied late, λ and $\tan \lambda$ are positive and the clock loses; if early, they are negative and it gains. Those are conclusions stated by Airy in one of his examples. What is more opaque in Airy's formulation, however, is that the actual amount of the escapement error is inversely proportional to the Q of the resonator.

Until the subject is further developed in the next chapter, a single numerical example must suffice. Suppose the driving force is consistently late by 45° in phase angle, presumably not as the result of a fault in the mechanism but by some design idiosyncrasy. The tangent of 45° is 1, and if we assume a Q of 10 000, equation (10) tells us that the clock will go slower by a fraction 1/20 000. In other words, it will lose by one 20 000th of a second per second. Steadily maintained, this amounts to a loss of 4.32 seconds per day.

It is tempting to argue (as Airy did) that clock escapements should be designed to apply their force with zero lag, as this makes the escapement error zero, but escapement error does no harm so long as it never varies. All unvarying 'errors' in a clock's rate are of course regulated out of existence; we can never observe them. Timekeeping only varies if the so-called errors should themselves vary. Horologists know this full well, and yet often seem to adopt a somewhat ambiguous attitude. In the back of the mind, a little voice tries to argue that an error which does not exist is safely out of the way. Unfortunately, being zero is not the same thing as ceasing to exist. Nothing is perfect, and an escapement designed for zero phase error may still wander a little one way or the other and so become responsible for random errors of timekeeping.

In a self-sustained oscillator such as a clock, the drive is triggered by the resonator itself, so ensuring – on average – a constant phase relationship with the pendulum. This is what we have been assuming, but the theory might just as well have been describing what are known as *forced vibrations*, in which a driving force of constant frequency is imposed from outside without sensing the phase of the resonator at all.

A forced oscillation is one that responds to some external driving force which need not be at the natural frequency of the resonator itself. When a soprano sings into a wine glass, the glass does not respond with its own note, but tries to vibrate at the singer's frequency. Forced oscillations of a resonant system are a subject of study for every student of physics, not perhaps in terms of wine glasses, which are less than ideal as pieces of laboratory equipment, but at least in terms of simple one-dimensional resonators. The only conclusion from such studies that matters here is this: whilst the applied force and the induced velocity of the resonator are necessarily at the same frequency, they are not necessarily in phase with one other. Phase agreement occurs only when the force is spot on the resonant frequency. At too high a frequency, the inertia of the resonator causes its velocity to lag in phase; at too low a frequency, the velocity is ahead. This is because the influence of inertia is smaller at the lower frequency; at an extremely low frequency it is only the *position* of the resonant element that is affected by the force, and this lags behind the velocity, as we have already seen in Figure 8.4, by 90° in phase. A discrepancy in phase between force and velocity can be imagined as an expression of the resonator's reluctance to vibrate at other than its own natural frequency. A more obvious expression of this reluctance is its much reduced amplitude of vibration. The wine glass will not shatter unless the singer hits very closely on the natural frequency of the glass and maintains it for long enough for the amplitude to build up and exceed the elastic limit of the material. I have never seen the trick performed, but it must demand a glass with a high Q and a singer with much less vibrato than is usual at the opera.

Very few clocks are required to work like wine glasses, though in the nineteenth century there were some whose pendulums were forced to vibrate in sympathy with a remote master pendulum. This was the earliest method of synchronizing clocks, invented – according to Hope-Jones (1949) – by the railway station master at Chester in 1857. By contrast, a clock that keeps its own time makes no call on an external frequency supply, the frequency of the drive being taken from the pendulum itself. This raises an interesting question, for if a wine glass can vibrate albeit reluctantly at a frequency that is slightly off-tune, there does not at first appear to be any corresponding reason for a clock pendulum to do so. The theory outlined in this chapter answers the question; the pendulum's frequency depends to some extent on the phasing of the drive in relation to its velocity.

A forced oscillation can thus be seen to differ from a self-sustaining oscillation by an interchange of cause and effect. The singer who is off-tune is the cause of a phase discrepancy. In a clock, a phase discrepancy on the part of the escapement is the cause of the pendulum's swinging slightly off-tune. The tangent rule applies equally well to the sympathetic vibrations of the station master's slave clocks as to any escapement error in his master clock. It is one of many examples in which physics seems oblivious to which is the cause and which the effect.

CHAPTER NINE

The phase circle

It is of little consequence that the sinusoidal driving forces discussed in the previous chapter are never encountered in mechanical clockwork. Any repeating pattern of force applied to a simple high Q resonator at or near its resonant frequency looks to it like a sinusoid; it can hardly tell the difference, and reacts just as it would to a sinusoid. This little known fact makes the calculation of escapement error in practical situations relatively straightforward. The method is as follows:

- estimate the time-varying pattern of force which the escapement applies to the pendulum;
- line it up with a sinusoid;
- compare the phase of the sinusoid with that of the pendulum's velocity to find the lead or lag; and
- apply the tangent rule given at equation (10) of the previous chapter.

Notice that the forces are to be thought of as depending on time rather than on position. In mechanical clockwork, this is the only thing that may make the procedure tricky, especially with recoil escapements.

The central role of the sinusoid in vibration theory is inescapable, and is made easier to understand in the light of a mathematical series named after J. B. J. Fourier, which tells us that any repeating pattern in time can be uniquely and completely analysed into a fundamental sinusoid and a set of sinusoidal harmonics at two, three, four, etc., times the fundamental frequency. The pattern of the driving force from an escapement is seldom sinusoidal, but its fundamental component is so, by definition. A pendulum cannot respond to any of the harmonics because it is virtually incapable of vibrating at any of the

THE PHASE CIRCLE

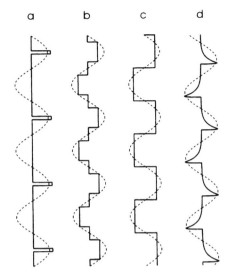

FIGURE 9.1
Force patterns of various escapements. Force is plotted horizontally and time vertically.

harmonic frequencies. This useful property is only true for high Q resonators with a single degree of freedom, such as pendulums and balance wheels. For a pendulum, therefore, we do not have to bother with any of the higher frequency components in the driving force; we need only pick out the fundamental. In general, this calls for the use of the calculus, but it can often be done either by commonsense or by eye.

A sinusoidal waveform is completely specified by an amplitude and a phase. Those are the only things we can know about the fundamental component of a repeating impulse pattern, and if we know them, we have all we need. In fact, we have more than we need. When escapement theory is formulated for steady state oscillation, as in the previous chapter, it turns out that the phase alone is sufficient, which in simple cases is obvious by symmetry, as the drawings in Figure 9.1 should make clear.

These graphs are plots of the lateral force applied to the pendulum, with time progressing vertically downwards. If the force repeats regularly and possesses the same symmetry as a sinusoid, there is only one possibility: the humps of the sinusoid must line up with the centres of symmetry of the impulses. The pattern Figure 9.1a could be that of a chronometer's single-beat escapement, with its sharp impulses shown acting towards the right in the drawing. That at Figure 9.1b, consisting of longer impulses alternately right and left, is typical of a pendulum clock with a Graham dead-beat escapement, ignoring the frictional drag forces. The pattern at Figure 9.1c could be the force exerted by a double-beat gravity escapement, in which one arm or the other is always in contact with the pendulum rod. A similar pattern could of course represent any recoil escapement in which a force is continually being applied in one direction or the other, ignoring drop and ignoring any variation there might be within the impulse itself. More problematical is the pattern at Figure 9.1d, but a simple escapement which can produce a gradually increasing force with a sudden fall to zero is discussed later in this chapter.

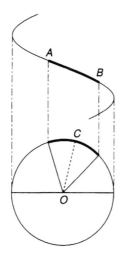

FIGURE 9.2
Plotting an impulse from A to B on a phase circle and finding its time or phase centre C.

When there is time symmetry, the next step is to estimate the phase lag λ relative to the moment at which the pendulum passes through its central position. (This is the same λ as that in the previous chapter.) Phase is a timing measurement, and is not one a clockmaker can make directly by measuring mechanical parts. However, if the points at which the impulse starts and finishes are known geometrically, and an amplitude of swing can be assumed, the phase lag is easily determined. One way of finding it involves drawing a sinusoid to represent the position of the pendulum versus time as in Figure 9.2 (top), but an easier way is to draw a phase circle with diameter equal to the pendulum's amplitude of swing, Figure 9.2 (below). The thickened section *AB* of the sinusoid is supposed to show the whereabouts of the impulse in the swing. On the phase circle the motion of the pendulum is represented by a point moving clockwise round the circle at a constant angular rate.

From knowledge of its position in the swing, which is horizontal distance on the diagram, the impulse can be marked out on the circumference of the circle, and its duration is then represented by the angle subtended at *O*. The radius *OC* bisects this angle, so *C* is the mid-point of the impulse in time (but *not* in space, as will be seen more clearly in the next example). The angular tilt of *OC*, measured clockwise from the vertical, is the phase lag of the escapement; this is the angle λ needed for the tangent rule given in equation (10) of the previous chapter. In the drawing, the tilt happens to be 13.5°. With a *Q* of 1000 (a modest clock), the escapement error would be about ten seconds a day losing.

A recoil escapement gives a more interesting picture, Figure 9.3. The impulse starts at *A*, pushing towards the right, hindering the pendulum until it turns back, and ends at *B*. It has occupied half a cycle or period, after which a double-beat escapement such as the grasshopper would start receiving its leftward impulse for the remaining half-cycle. The impulse occupies half of the phase circle, and is bisected by the radius *OC*. The angular tilt of *OC*, here about −37° (measured clockwise from the vertical) is the phase lag λ. As with all recoil escapements, the time centre of the impulse is reached before the pendulum

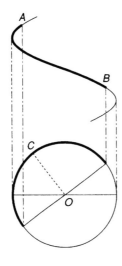

FIGURE 9.3
Recoil impulse from A to B mapped on to the phase circle, with its phase centre marked at C.

passes through its centre of swing, which makes λ negative, making the 'fractional loss of time' negative also, from the tangent rule. In other words, the recoil escapement causes the pendulum to gain, here by about 33 seconds per day, taking the same figure of 1000 for Q.

A constant escapement error is of no consequence, but it is unlikely to stay constant if the pendulum's arc of swing should vary. This effect is most marked for a recoil escapement, Figure 9.4. An increased amplitude of swing, produced perhaps by an increase of driving force, increases the radius of the phase circle. The end-points of the impulse, A and B, will not move in space, i.e. laterally on the diagram, because they are fixed by the geometry of the clock's mechanism, but the corresponding phase angles will change. In moving from C to C', the phase of the impulse is advanced, and the gain from escapement error therefore increased. To give a sense of scale, the 10% increase of arc shown in the drawing would need about 20% more driving force at the escapement; the phase 'lag', originally −37°, then becomes −43°. Assuming a Q value of 1000 again, the formula

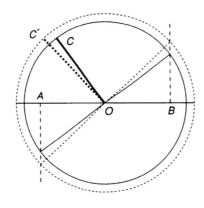

FIGURE 9.4
Change of phase centre resulting from increased amplitude with a recoil escapement.

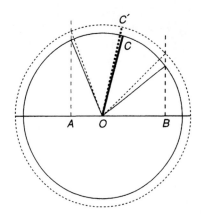

FIGURE 9.5
Comparatively small change of phase centre from increased amplitude with a non-recoiling escapement.

$(\tan \lambda)/(2Q)$ can be evaluated for the two cases, giving an increase of fractional error from 0.000 38 to 0.000 47, which is a gain of about 8 seconds a day. Notice that I am assuming Q to remain unchanged, which would be a reasonable assumption if the increased arc of swing had been due only to a change in the clock's driving force, caused for example by rewinding a mainspring.

It is interesting to repeat this exercise for a lagging impulse with no recoil. Most non-recoiling escapements seem to have a positive lag and, for such escapements, an increase in driving force can be seen to advance the phase of the impulse just as for a recoil escapement, though to a very much lesser extent, Figure 9.5. The time-centre of the recoil impulse was to the left of centre on the phase diagram and moved farther to the left. The time-centre for the lagging escapement is to the right of centre, but it also moves left with increasing amplitude. We thus have the interesting result that an increase of driving force makes a recoil escapement's gaining error increase but makes a non-recoiling escapement's losing error decrease. In both cases, the upshot is a gain. Note in passing that central impulses give no escapement error, and are inherently insensitive to changes of

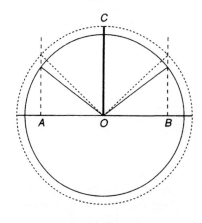

FIGURE 9.6
Central impulse is insensitive to changes of amplitude.

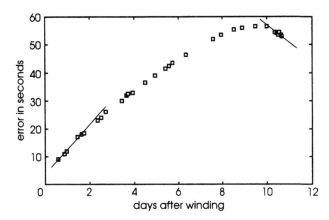

FIGURE 9.7 *Cumulative time error of a spring driven clock with Brocot escapement.*

amplitude, Figure 9.6. Surrounded by marginal gains on both sides, the sensitivity of the central impulse to changes of arc is not only zero but stationary, which makes it all the more desirable for use with isochronous resonators.

Because of circular error, pendulums are not isochronous. Any variation of amplitude alters the circular error as well as the escapement error, which brings us to the place where many horologists are liable to slip up. Circular error is a losing error and the escapement error from a recoil escapement is a gaining error; in a recoil escapement, therefore, the two errors tend to cancel, whereas with the lagging impulse of a dead-beat escapement, both errors are losing and add up. Whilst each of these statements is true, it is too easy to forget that we are only concerned with marginal variations. The popular French escapement designed by Achille Brocot was to provide me with an outstanding example in which the errors do indeed reinforce whilst their marginal variations cancel.

The modern French 'four-glass' clock I tested was larger than a travelling clock, and had a miniature mercurial pendulum. However, it was spring driven and because it lacked any special mechanism such as a fusee to compensate for loss of torque as the spring unwound, the pendulum's amplitude diminished steadily through the week. After lubrication, this particular clock would run for ten days on one winding, during which period the amplitude fell from about 5° to about 3°. The graph in Figure 9.7 shows the time error at various odd times throughout the ten days. During the first seven days, it had gained approximately 42 s, a figure of absolutely no significance, for the clock was later regulated to keep closely to time when wound weekly. Here I am only concerned with the departure of the graph from linearity. A perfect clock gaining 42 s in 7 days would do it at a uniform rate of 6 s a day and a graph of its time error would be a straight line. The clock under test gained 25 s in the first half week and 17 s in the second. If regulated to eliminate the gain of 42 s, it is easy to calculate that it would gain 4 s in the first half of the week and lose them in the second. Nobody could reasonably complain about such a slight variation through the week by a clock with no seconds hand. Indeed, the owner had been delighted with its performance, and I was interested to find out why the timekeeping was so good.

I have drawn two tangents to an imaginary smooth curve through the observed points in order to estimate the clock's rate at the start and end of the test. They indicate rates of +8.3 s/day and −5.5 s/day for the amplitudes 5° and 3° respectively, a decrease in rate of about 14 s/day. As the Brocot is not a recoil escapement, I had not expected there to be a large escapement error, and the fact that the clock went considerably slower at smaller arcs of swing puzzled me at first. Circular error, which is easy to calculate, has the opposite effect. The formula for fractional circular error is $\alpha^2/16$ losing, where α is the amplitude of swing in radians; multiplication by 86 400 converts it to a loss in seconds per day. The results of this calculation for amplitudes of 5° and 3° are −41 s/day and −15 s/day respectively. From variation of circular error, therefore, the clock would have been expected to go 26 s/day faster at the end of the experiment, not 14 s/day slower! It was a huge discrepancy.

The explanation could only be sought in the variation of escapement error with amplitude. The Brocot escapement, Figure 9.8, has D-shaped cylindrical pallets, often of semi-precious stone, though usually of steel when the escapement is not on show! It is an interesting escapement for several reasons. It has a dead-beat action with the faces of the escape wheel teeth acting as the dead surfaces, in contrast to the Graham escapement where the dead surfaces are on the pallets. Impulse is delivered as the point of the escape wheel tooth slides down the acting semi-circular half of the pallet jewel, with its ever increasing slope. The resulting force on the pendulum is therefore far from constant during the impulse, for as the tooth nears the end of the pallet, it has greater leverage. The strongest part of the impulse is delivered well after the pendulum has passed through its centre of swing. I was already familiar with lopsided impulses of this kind from the Hope-Jones arrangement shown earlier in Figures 4.1 and 4.2. This would be even more

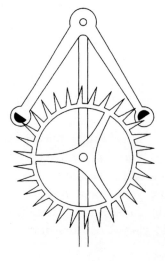

FIGURE 9.8
Brocot escapement.

Rates in seconds/day	circular error	escapement error	total
end of test (amplitude 3°)	−15	−65	−80
start of test (amplitude 5°)	−41	−25	−66
difference	26	−40	−14

Table 9.1 Theoretical error calculations for Brocot escapement

pronounced with the Brocot escapement because the early part of the impulse would be largely counteracted by pallet friction. Escapement error might therefore be much more important than I had imagined.

As I showed in the previous chapter, it is not difficult to calculate escapement error provided that we know two things, the Q of the pendulum and the phase lag, λ, of the impulse. I knew neither, and by the time I had become really interested in having them, the clock had been returned to its owner. They would just have to be estimated. The Q of a heavy pendulum is upwards of 10 000. This was a much lighter pendulum, with miniature sealed mercury jars in a brass frame. The ornamental nature of the bob's shape and its comparatively large arc of swing would undoubtedly add to the air resistance. As a very rough estimate, I put the Q at 1000. As for the phase lag, I would first have to guess at the geometric centre of the strongly biased impulse. I knew that the wheel tooth escaped from the pallet stone at a semi-arc of 3°, for that was when the clock stopped, so I took the centre of the impulse to be at 2.5°.

The phase lag of an impulse concentrated at an arc angle 2.5° past the centre would be $\lambda = \arcsin(2.5/\alpha)$, where α is the amplitude of swing in degrees. This calculation can be followed from Figure 8.3 in the previous chapter, which is marked out appropriately. Using the tangent rule at equation (10) in that chapter the losing escapement error is

$$\frac{1}{2Q}\tan\left\{\arcsin\left(\frac{2.5}{\alpha}\right)\right\}$$

A scientific calculator deals with the division, the arcsin, and the tangent in one step apiece. The theoretical balance sheet for the clock under test is shown in Table 9.1. The experimental figures were −5.5 at the end of the test and +8.3 at the start. To the nearest whole number, the difference was −14 seconds per day! Bearing in mind that Q and λ were not measured but only guessed, there is an element of luck here, but the agreement gave me sufficient confidence to proceed further.

The escapement error more than wipes out the circular error when taken over the whole run, but this clock was not meant to go for more than seven days on one winding. The first seven days worked out more favourably. It is easy to calculate the rate deviations for circular and escapement error for different arcs of swing, and continuous graphs covering the range 3° to 6° are shown in Figure 9.9. When the two deviations are added together,

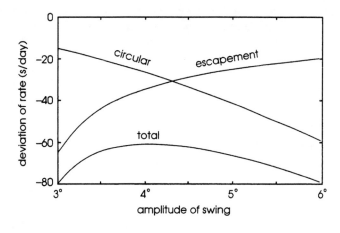

FIGURE 9.9
Cancellation of circular error and escapement error variations.

we see that the clock could well have been designed to work at or around the amplitude of 4°, where the total rate deviation is flat. The cancellation of variation goes far towards explaining the constancy of this little clock, which, after correct regulation, keeps time to within a few seconds a week in spite of considerable mainspring variation. From now onwards, I shall look at Brocot escapements in an entirely fresh light.

It is worth repeating that we have here large losing contributions to the rate from circular error and escapement error, together totalling about a minute a day, and yet they cancel out!

To obtain the escapement error for an unsymmetrical impulse, one has to find its phase centre. The example in Figure 9.1d is intended to represent the type of impulse with which we may have to deal. The phase of the fundamental frequency component cannot be determined exactly without the calculus, but I have succeeded in finding a construction which should enable any enthusiast to make a reasonable estimate without recourse to calculation. The method is given without proof, as a fuller account can be found in the *Horological Journal* (Woodward 1989).

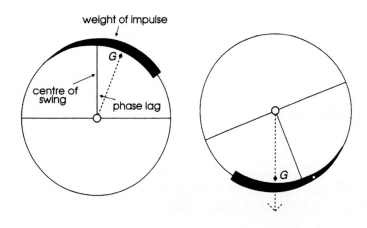

FIGURE 9.10
Method of estimating phase centre of an unsymmetrical impulse.

The method is basically mechanical. What we must do is to think of the phase circle as a wheel, Figure 9.10. The rim is weighted in proportion to the impulse at the appropriate phase angles. The angle of the radius through the centre of gravity G, measured clockwise from the centre of swing, is the required phase lag. This angle could, in principle, be found experimentally. Having loaded the wheel, it would be released and would settle with the centre of gravity vertically below the centre. Nobody would think it worthwhile to carry out such an experiment in reality, but mechanical intuition can be a remarkably accurate substitute.

CHAPTER TEN

The Shortt free pendulum

There is no such thing as a free pendulum, because a pendulum can never be totally isolated from its environment. If it could, its Q would be infinite and we would have perpetual motion. Freedom has always been an unattainable ideal, but the degree to which it has been approached is truly remarkable, so much so that the term *free pendulum* need hardly be regarded as a contradiction in terms. An earlier concept was that of *detachment*, which is more easily defined.

Detached escapements were first contrived for the balance and hairspring resonators used in watches. After receiving an impulse in mid-swing, the resonator detaches itself from the escapement which promptly goes to sleep. The balance swings freely out to its extreme arc and back towards the centre, where it wakes up the escapement, takes an impulse and detaches itself once again. This deserves a closer look.

In a watch with a detached lever escapement, the main arbor of the resonator is called the *balance staff*. It carries the balance, which is the inertial element usually shaped like a wheel with its mass concentrated in the rim, and also has attached to it one end of the hairspring, the other end of which is (in effect) fixed to the frame of the watch, Figure 10.1. Also mounted on the balance staff is a fixed disc, known picturesquely as a *roller*, on which is mounted a pin of D-shaped section. This pin is the point of contact between the escapement and the resonator, acting as a kind of crank handle for the balance. During the extreme of swing, the balance is free, but as it passes through its central position, the impulse pin finds a small obstruction in its path, Figure 10.2. It is one arm of a fork on the end of a lever connected to the escapement. The pin pushes the fork out of its way, so unlocking the escapement, which causes the pin to be impulsed from behind by the other arm of the fork. Having to unlock the escapement is a small price to pay for detachment.

Pendulum clocks for the most part by-passed this development. Because of their inherently higher Q, escapement friction had never been as significant; weight-driven

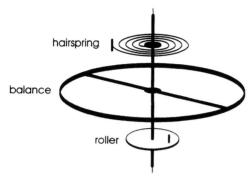

FIGURE 10.1
Resonator of a mechanical watch in diagrammatic form. The peg mounted on the roller connects with the lever escapement.

FIGURE 10.2
One form of lever escapement for a mechanical watch.

pendulum clocks already kept good time with undetached escapements. Also, detachment is more difficult to arrange for a pendulum because of its smaller arc of swing, though various detached escapements have been invented, including one by Sir George Airy made by Dent for use at the Royal Observatory. It is far from clear that detachment improved the timekeeping accuracy of clocks as dramatically as it had done for watches, if indeed at all. Complete freedom, however, would be different.

A free pendulum has to avoid even the small chore of ringing for its tea. It requires that the tray be brought in unbidden at the appointed time! This sounds like a logical impossibility, for there would then have to be a clock in the servants' hall, but that was precisely how the trick was performed. Leaving aside for a moment the seemingly circular nature of the argument, one can see that a pendulum not having to call for its tea might be allowed to go on swinging freely for a considerable time, for servants who can judge an interval of a second can just as well judge an interval of a minute and still catch their master at more or less the right moment. The *right* time for the master to take tea is something only he can judge, but the astute servant, always arriving safely early, watches over his master and judges whether he had cut it too fine. If so, he simply puts the kitchen clock forward a little. This may seem a roundabout way of ensuring that a pendulum receives its impulse at just the right phase in just the right swing, but it is the system Shortt used, and it worked.

The free pendulum clock designed by William Hamilton Shortt dates from 1922. Although named after Shortt, the clock is closely coupled with the name of Hope-Jones, whose Synchronome workshops in Clerkenwell were placed at Shortt's disposal and eventually used for manufacture of the clock in quantity. It must have been an interesting collaboration, for Hope-Jones seems to have been something of a back-seat driver during the development phase. Certainly, the Shortt clock owed much to the early form of Synchronome described in chapter 4, with its gravity impulse acting through a roller and being reset by an electromagnet. As a result, the Shortt is usually classified as an electric clock, but this can be misleading, as electricity was not used for any operation crucial to the accuracy of its timekeeping. For impulsing their pendulums, the leading European makers of the day, Shortt in England, Leroy in France and Riefler in Germany, all shunned the use of electromagnetic impulsing, preferring to rely on pure mechanics. Although electromagnetic impulsing had been used by Alexander Bain as early as 1843, useful components such as amplifiers – which we now build into wrist watches! – were still clumsy and unreliable heaps in 1922.

Shortt's clock differed from its two European competitors in various respects. All enclosed their clocks in sealed cases to isolate the pendulum from variations of atmospheric pressure, but Shortt used a high vacuum whilst the others worked at pressures not far below atmospheric. The impulsing arrangements were quite different. Leroy's pendulum received recoiling impulses every second from springs operating in a similar way to gravity arms. The Riefler escapement supplied the pendulum with recoil impulses too, ingeniously by rocking the suspension spring on knife-edges, which left the pendulum looking completely free though in reality it was not even detached. Only the Shortt clock could boast a free pendulum impulsed centrally by a gravity arm at half-minute intervals. Although not a clock for a domestic environment, it had a functional beauty of its own, Figure 10.3, and can be admired in the setting of the Old Royal Observatory at Greenwich or in the new Space and Time Gallery at Liverpool Museum.

For a pendulum *in vacuo*, impulsing is a matter of the utmost delicacy, because the energy requirements are several times smaller than at atmospheric pressure. In the Shortt clock, the gravity arm operates on a tiny roller wheel 6 mm in diameter, pivoted in an elaborate looking bracket on the rod, Figure 10.4. There is no count wheel to drive, and for twenty-nine out of thirty swings the pendulum is on its own. On the thirtieth swing, and in the very nick of time, an external timer unlatches the gravity arm, which drops on to the roller in the course of the pendulum's swing to the left. The gravity arm falls away from the rim of the roller as the roller recedes, so giving impulse, and is reset before the pendulum returns.

The complexity of the bracket is partly to facilitate fine adjustment, and partly for safety, the two purposes being neatly combined. The bracket is located by screw-points for adjustment vertically and laterally, and the whole thing 'gives' if, through failure of the gravity arm to reset, a head-on collision should occur between the gravity arm and the roller. Without such give, one dare not think what damage would result from a collision between a wheel made with the delicacy of a watch escape wheel and a pendulum weighing 7 kg or more assisted by the leverage of the rod!

The action of the Shortt movement can be followed from the three drawings in Figure 10.5. The gravity arm is held clear of the impulse roller by a hemi-cylindrical locking stone L which is moved out of the way when the external timer energizes the solenoid A. (The locking piece is friction tight on its pivot and stays wherever it is put.) The impulsing jewel also is hemi-cylindrical in shape, and its flat surface falls on the wheel. After a short

THE SHORTT FREE PENDULUM

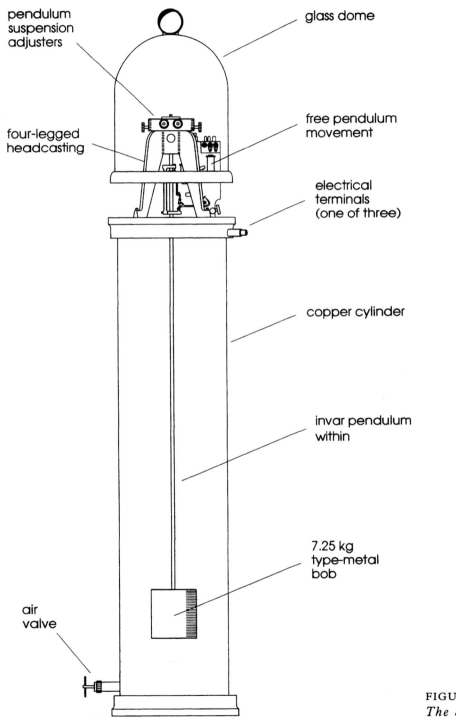

FIGURE 10.3
The Shortt clock.

This Shortt clock, made in 1937, was acquired by the Liverpool Museum in 1993. The paired dials on the Synchronome slave are operated by the resetting circuits for the gravity arms of the master and slave pendulums respectively. As these two actions take place in rapid succession, the dials should be in exact agreement for all but a fraction of a second every half a minute. The large dial includes a seconds indication, driven by an auxiliary device on the slave pendulum to provide an audible seconds beat for astronomical and other purposes.

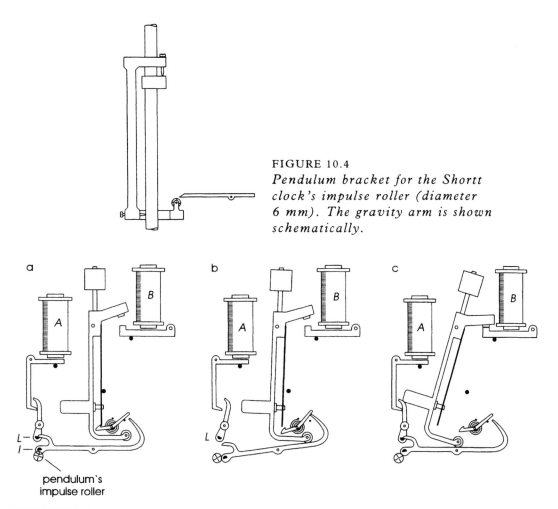

FIGURE 10.4
Pendulum bracket for the Shortt clock's impulse roller (diameter 6 mm). The gravity arm is shown schematically.

FIGURE 10.5
Three stages in the action of the Shortt movement. (a) Rest position with wheel clear of the impulse pallet I. (b) Solenoid A has released the locking pallet L, the gravity arm has impulsed the pendulum, and the tail of the arm has released the catch holding the resetting lever, which is about to reset the gravity arm. (c) After completing the lift of the gravity arm, the resetting lever has reset the locking pallet and has closed the electrical contact, enabling solenoid B to return the resetting arm to its catch.

dead roll, the sharp corner pushes on the receding wheel as it drops, so impulsing the pendulum.

In an ordinary Synchronome clock, the gravity arm is reset by a magnet, but Shortt considered this too brutal. Instead, the tail of the arm releases a spring catch holding a sturdier resetting lever, whose lower end is now free to roll along the delicate gravity arm with a cam-like action and reset it gently. The resetting lever finishes by resetting the latch that held the gravity arm, finally making contact with the armature of a second magnet B,

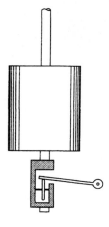

FIGURE 10.6
Early form of impulsing by Shortt underneath the bob.

which resets the lever. The final jolt is softened by a leaf spring fixed to the lever. These three drawings were prepared by computer, enabling each stage of the process to be simulated by a slow form of animation. In preparing such drawings, one learns much about the subtlety of the clock, and of Shortt's especial forte for mechanical design.

The crux of the Shortt clock is, of course, the method by which the pulse of current for releasing the gravity arm is made to arrive at just the right moment. It is instructive to learn from Shortt himself the stages through which the design passed towards this end. We are fortunate to know something about them from a talk he gave (Shortt 1929) to the British Horological Institute describing the various mechanisms he had 'constructed, improved, re-constructed, and scrapped' during his association with Frank Hope-Jones.

At first the impulse wheel was pivoted in a G-shaped bracket fixed underneath the pendulum bob. In the drawing, Figure 10.6, the bob is to be imagined swinging in and out of the page. As it passes through the central position in each direction, a tab fixed to the bottom of the 'G' actuates a rocker arm not shown in the drawing. This unlatches the gravity arm in time for it to fall on the wheel and deliver impulse in each direction of swing. The arm was reset by an electromagnet without the benefit of a separate resetting lever. It was a somewhat ungainly arrangement, and although the pendulum was detached, it was in no sense free. As already explained, a free pendulum is not allowed to unlatch anything at all; indeed, Shortt himself stated that more than half of the energy given to the pendulum by the gravity arm was used up in releasing the arm for the next impulse!

For some time, and to Hope-Jones's evident disgust, Shortt persisted with impulsing every second, being convinced that astronomers would demand a clear beat every second when observing star transits. This may have been a fortunate circumstance. The gravity arm had to be so delicate as to cause Shortt to introduce a separate rolling contact lever of the type already described, and the resulting slow reset, taking about three quarters of a second from start to finish, is what initiated the train of thought that led to the freeing of the pendulum.

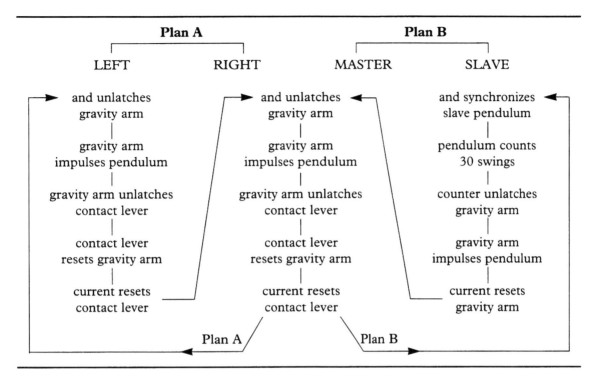

Table 10.1 Evolution of Shortt clock design

Still impulsing every second, by the time the resetting was complete it was almost time to initiate the impulse for the next swing, which gave Shortt the idea of eliminating the unlatching altogether by using two separate escapements in duplex, one for impulsing to the left and one for impulsing to the right. Each escapement had its own gravity arm and its own resetting contact lever, the resetting current for the contact lever of one escapement releasing the gravity arm of the other escapement. This was simply a matter of putting the two relevant electromagnets in the same series circuit. The pendulum was now technically free, though only for a very limited proportion of the time.

The duplex action is summarized in what I have called Plan A, which is represented diagrammatically in Table 10.1. It worked, though not well. Because of impulsing on every swing, the forces were still too small for comfort, and the mechanism was edgy. At length Shortt was forced to consider impulsing at wider intervals as Hope-Jones had advocated all along. To time a longer interval, the duplex idea would no longer serve; a proper timer would be needed if the free pendulum's gravity arm was to be released only once in thirty seconds. For such a timer, the obvious candidate was a standard Synchronome master clock, now to become the slave for a higher master. This was to be the final design. We can see from Plan B that the functions of the master pendulum are exactly the same as before, but the slave is necessarily different, as it has the task of counting out the 30-second intervals.

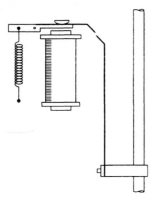

FIGURE 10.7
Shortt's hit-or-miss synchronizer fitted to the rod of the Synchronome slave clock.

The one remaining problem was to keep the slave pendulum synchronized with that of the master. If this could be done by readjusting its phase every half a minute, all well and good. A clock like the Synchronome, with its very heavy pendulum, cannot easily be stopped and restarted like a modern count-down timer (which was precisely the function needed). Fortunately, only a tiny adjustment is called for, as the time kept by a Synchronome is almost as good as that of the free pendulum itself. The trick was to regulate the slaved Synchronome such that it would never overtake the master pendulum, but always run a little slower. Then, on receipt of current from the master's solenoid B (see Figure 10.5), the time of the slave was compared with that of the master, and the phase of the slave pendulum given a small hastening boost if necessary. This was accomplished by a leaf spring fitted to the slave pendulum rod.

Shortt's *hit-or-miss synchronizer* shown in Figure 10.7 is admirably simple in operation. The current from the master energizes an electromagnet which draws down a prong. If the slave pendulum is slightly late on its swing to the left, the prong is caught under the leaf spring for half a period (a *hit*), otherwise not at all (a *miss*). When deflected by a hit, the spring adds to the pendulum's restoring force and hastens it by a small fraction of a second. The actual amount of hastening is chosen to be twice the natural loss of the slave pendulum in half a minute, making hits necessary on about half the occasions. In practice, hits and misses alternate with remarkable monotony if the system is properly adjusted. The free pendulum only shows its mettle when the alternation is occasionally broken. Great patience is called for as one waits for such occurrences, and even when they do occur it is more than likely that they are caused by imperfect rating of the slave rather than its lack of regularity.

An engineer's natural tendency is to view the hit-or-miss principle with slight suspicion. A clear hit or a clear miss is all well and good, but what about the inevitable borderline cases? With sharpened points, they are unlikely to happen, but the beauty of this synchronizer is that an occasional sitting on the fence does no harm whatever. The prongs may perhaps lodge edge to edge, as fingernails may be persuaded to do, and slip off at some indeterminate future moment, but if it happens only occasionally it is not of the slightest consequence.

The Shortt clock turned out to be a great success; in the years between the two world wars, it was adjudged by many to be the most accurate clock there was. By 1956, the Synchronome Company had made just about one hundred and exported them to observatories the world over. The Royal Observatory at Greenwich had several, though to judge from the list given by Frank Hope-Jones, France was in no hurry to purchase a clock made in England. This is understandable, as the French possessed an excellent contender in the clock by Leroy of Paris. Whether or not it was fully a match for the Shortt is hard to judge. Over long periods, the variation in the going rate of a Shortt clock might be as little as a second or two a year, which is probably close to the limit of what is possible for any clock whose resonant frequency depends on the uncertain force of Earth's gravity.

Nobody has yet succeeded in showing which feature of the Shortt clock contributed most to its success, and I like to think that all played an equal part. The clock was meant to be run in a thermostatically controlled environment, notwithstanding its invar pendulum rod. The vacuum boosted its Q to about 110 000. Some say that this was sufficient explanation in itself, but Rawlings (1993, pp. 95–98) has argued that it might have been a disadvantage. No benefit can be reaped from a high value of Q unless the much reduced impulses can be kept constant in proportion to their much reduced strength. If the *absolute* variation in the force of the impulse remains constant, a higher Q increases the variation of amplitude and hence that of circular error. The mechanics of the Shortt escapement is of exquisite delicacy, but it is an open question as to whether the impulsing was still too frequent, even at half-minute intervals.

In the years between the two world wars, the Shortt clock was used as a primary standard at the Royal Observatory, Greenwich. It was a fitting climax to the era of the pendulum.

CHAPTER ELEVEN

Aiming too high

In the end it was to prove fortunate that my first attempt at a free pendulum ended in failure, for otherwise I would not have been spurred on to make W5. W4 was a compact wall clock, about the size of a Vienna regulator, but with two pendulums of different lengths swinging side by side. The shorter pendulum, a half-second slave, operated a pair of count wheels and took impulse every fifteen seconds. The longer one was the master, which swung freely for four minutes before receiving an impulse, during which time it had made 173 full vibrations. I took some pleasure in choosing that prime number, but almost any other number would have done equally well, for if there is no contact with the free pendulum between impulses, it can hardly matter how many swings it has made. Nothing in the mechanism depends on the particular number, though it must of course be an integer. The bob could be raised to give 174 vibrations in the four minutes and work equally well without any other adjustment. When details of W4 were prematurely published (Woodward 1974), an unwise step in any branch of research, the irrelevance of the master pendulum's frequency did indeed puzzle some readers, though not for long.

The idea of mounting slave and master pendulums side by side in one cabinet immediately makes a physicist think of possible interactions, for it is well known that each can affect the other, and even alter its frequency of vibration. In a master and slave arrangement there must be one-way interaction only, the master controlling the slave via the hit-or-miss synchronizer. If the slave were to influence the master, the whole concept would be jeopardized. Simple experiments showing how pendulums can interact are easily rigged up, but the most remarkable was that carried out by Brown and Brouwer (1931) with three Shortt clocks. This experiment has become a horological classic.

Three clocks, SH20, SH21 and SH22, were mounted on independent massive masonry piers on solid rock foundations with their pendulums swinging in different planes 120° apart, and each clock's rate was found to be modulated by the difference in frequency of the other two. It is hardly credible that the rocky foundations should have moved in response to the swings of a pendulum, but this was the only feasible mechanism for the interaction. It brings home the fact that, with a high Q resonator such as a Shortt pendulum *in vacuo*, every tiny possibility must be taken into consideration.

There is little danger of interaction between pendulums tuned to entirely different frequencies, and I had considered tuning my master pendulum to the usual 0.5 hertz, with the slave at 1.0 hertz. This was indeed suggested by Hope-Jones (1949, p. 223), but I was suspicious of frequencies so closely related harmonically. Such fears must surely have been groundless, because pendulums – unlike bells or quartz crystals – do not give off harmonics. Be that as it may, my chosen frequency of 0.720 833 hertz would undoubtedly be safe from interference by a pendulum of 0.500 000 hertz.

What happens if a Shortt master pendulum is mounted close to its slave when both pendulums, nominally of the same frequency, are swinging in the same plane, is something to puzzle over. The synchronizer ensures that the slave remains closely in phase with the master, but its *resonant* frequency is very slightly less than 0.5 hertz. Does this difference protect the master from interference, or does the operation of the synchronizer destroy that protection? Here is a nice exercise for an expert in the theory of coupled oscillators. I have never seen it explicitly stated that the slave of a Shortt clock should be mounted far away from the master, nor that the two pendulums should swing in orthogonal planes.

In a nutshell, my doomed free pendulum clock W4 worked as follows. The two pendulums each had intermittent grasshopper escapements of the type described in chapter 6, driven independently from separate trains, except that the master pendulum had no count wheel to drive. Had I not tried to be too clever by impulsing the two pendulums at different intervals, I believe W4 might have been a success. As things were, the four-minute interval was counted out by a vernier arrangement using a single pawl on the slave pendulum to drive two count wheels simultaneously, one of fifteen teeth and the other of sixteen. Each wheel had a pin in its rim, the fifteen-toothed wheel triggering the slave's impulse just as in W3. The two pins came into coincidence every four minutes, when the fifteen-toothed wheel had made sixteen revolutions and the sixteen-toothed wheel had made fifteen. This coincidence operated a mechanical *AND-gate*, a mechanism which produces an output when one pin *and* the other pin operate on it simultaneously. The output from the gate depressed the master pendulum's impulse hook in time for it to take impulse. When the recoiling impulse began, the gated detent on the master's escape wheel dropped, and in so doing operated a hit-or-miss synchronizer for the slave.

Lamentably, W4 would go for a few days or weeks and then stop. The longest run was five months, during which the timekeeping showed good promise. The fundamental flaw was the overloading of the slave pendulum. Even the simplest count wheel can consume as much energy as all the other pendulum losses combined. With two count wheels and the gating mechanism, air resistance was no longer the prime factor governing the slave's amplitude of swing, and plain friction was too variable a force to keep the amplitude inside limits which the mechanism could handle. The 1 kg free pendulum, on the other hand, needed astonishingly little power. Its escape wheel of thirty pins was driven by a single mesh of ratio 1:12 from a barrel half an inch in diameter, causing the weight of 340 g to drop a mere 280 mm in seven days with no pulley. The energy required to drive the lighter slave pendulum was seven or eight times greater, a fact that speaks for itself.

In desperation, I fitted a large fan to the slave pendulum, in an attempt to make air resistance the controlling factor. This necessitated a larger driving weight, until there was no room left in the case for any further addition. In all my work with electronic engineers, I have learned that measures of this kind are seldom successful. It is important to know when the design is wrong and make a fresh start.

Years passed before I gave any further thought to horological matters, during all of which time W4 pointed mockingly at twenty minutes past ten. As soon as I had retired from my professional work, pressure was applied to me to make the clock go or to scrap it. This set me thinking about imitating the Shortt clock, but without using electricity. Much of W4 could be salvaged, such as the dial, casework and pendulums, enabling me to concentrate on the mechanism alone. When the new clock – described at length in the next chapter – was finished two years later, one of my friends remarked that I had managed to get the wall clock going at long last. As it looked much the same on the surface, nothing I could say would persuade her that it was not the same old clock, which is what comes of putting new wine into old bottles.

CHAPTER TWELVE

W5

The Shortt clock was a product of the electromechanical age, which preceded the electronic revolution by a century. In 1920 the pendulum still held sway, but the electromagnet offered possibilities of control undreamed of in pure mechanics. Without such stimulus, the Shortt clock would never have evolved. However, with the benefit of hindsight, it struck me as an interesting challenge to see what might have been possible without making any use of electricity at all. The Shortt clock uses electricity to rewind the remontoire escapements of its free pendulum and its slave, which are both mechanical tasks. It uses electricity for communicating between the two pendulums, which might be more problematical. Finally, the electrical circuits make it a straightforward task to take off time signals for radio transmission or other purposes, a facility I would not need. My aim with W5 was to see how easy it would be to carry out the first two tasks mechanically.

The success of W5 proves nothing of any great consequence, for it was always obvious that simple electromechanical operations could be simulated mechanically if sufficient trouble were taken. For me, the question was whether the principle of the Shortt clock could be reproduced in a horologically *elegant* way, and I like to think that W5 has answered that question adequately. I am far from convinced that the same can be said of the design sketched by Hope-Jones in his book on *Electrical timekeeping*. There – unbelievably – the gravity arm is pivoted eccentrically on one of the spokes of the escape wheel of an ordinary regulator clock. I had seen this diagram long before embarking on W5, but had dismissed it from my mind as absurd.

In W5, as in the Shortt, both pendulums are impulsed at half-minute intervals. To fit the clock-case that had been made for W4 with its 173 vibrations in four minutes, the 'free' pendulum was lengthened a trifle to make twenty-one vibrations in thirty seconds. It is impulsed by a gravity arm and roller wheel just as in the real Shortt clock, except for one thing which I have never liked about the Shortt escapement. With a touch more

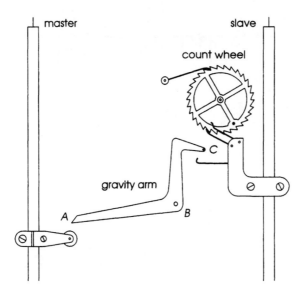

FIGURE 12.1
Outline plan of W5.

prejudice than science perhaps, I had been determined not to unlatch the gravity arm and allow it to fall freely on to the impulse roller. In my clock, the slave pendulum would lift the gravity arm off a rest and lower it gently on to the free pendulum's roller wheel.

The idea was to equip the slave pendulum with a hook for this purpose. An extract from my drawing of the final clock is shown in Figure 12.1. The slave pendulum on the right is equipped with a bracket in which three arms are pivoted, all of which are tail-heavy. The lower of the two arms at the top is the pushing pawl for the count wheel, and the top arm is a lever to sense the passage of a pin in the rim of that wheel. By a simple mechanism concealed within the bracket, this lever is connected with the hook at the bottom in such a way as to lift the hook when the lever is depressed. Once every thirty seconds, therefore, the hook rises and catches itself on a pin at C in the gravity arm, which is pivoted at B. Initially I had no idea how the gravity arm would be propped up before being grasped by the hook, but the plan was for the slave to lift the arm as it swung to the right, and on the return swing to lower it on to the master's roller. The touch-down would arrest the arm abruptly, but the hook would continue with the swing of the slave pendulum far enough to disengage itself immediately in true Harrisonian style.

Much remained to be thought out, but whatever had propped up the arm at the outset would obviously have to move out of the way before the arm could be lowered. Then, after impulsing the master pendulum, the arm would have to be reset. So pre-occupied had I been with the question of how all this might be accomplished that I had given no thought to how the slave itself would be impulsed. As a new design is being thought out, one problem at a time seems quite enough, but suddenly it dawned on me that two problems had been solved at once. To lift the gravity arm and park it at a lower point would in itself impulse the slave, and it would be a gravity impulse at that. This was a source of

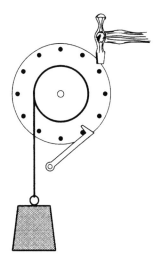

FIGURE 12.2
One way to release a detent!

great delight, as it meant that W5 would need only a single train and one escape wheel. The two pendulums would share one gravity arm, a trick I have not seen before.

The swan-like shape of the gravity arm is dictated by the positioning of the two pendulums. The line of the slave's hook is such as to form the most direct linkage between the slave pendulum and the gravity arm, at right angles to the line joining their respective pivot points. The underside of the gravity arm from A to B is likewise at right angles to a radius from the master pendulum's point of suspension to the point A when the arm is resting on the roller. This ensures a dead roll.

Once the gravity arm has fallen, it must be reset promptly enough to be out of the way by the time the roller returns. In the Shortt clock, the best part of one second is available, which gives time for the resetting to be performed at quite a gentle pace. With my shorter master pendulum, I had less time, and a heavier arm to reset. The work would be done by the escape wheel, which would have to be released from its detent by the free fall of the gravity arm. After the impulsing, the arm would have either to impact on the detent and knock it out of engagement, or – better still – jolt the wheel and so cause the detent to fall out of engagement by itself. This last idea, Figure 12.2, was much more appealing, as it would be yet another Harrisonian hop.

I can no longer retrace the line of thought that led me to the details of the final design of the remontoire shown in the series of drawings Figure 12.3, but I do remember taking a considerable time to work it out. It involves a linkage in the form of a triangular ear pivoted on the gravity arm. The ear acts as a prop to hold the gravity arm in its resting position when out of action. Notice that it rests on a pin of the escape wheel. The arbor of the detent carries another arm close to the clock-plate at a convenient depth for the banking pin. This is a detail, for the two arms move as one. The various stages of the action can be followed from the captions.

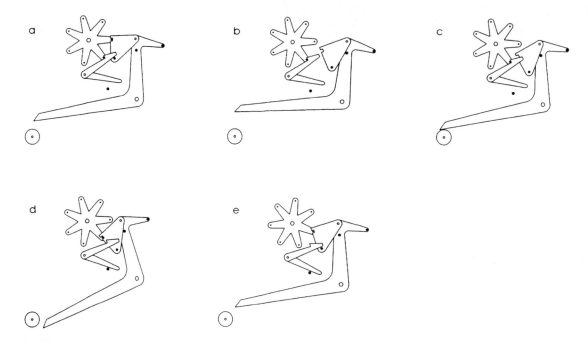

FIGURE 12.3
Five stages in the operation of the remontoire. (a) Rest position. (b) The slave has recoiled the arm and the ear has fallen to its banking. (c) The slave has lowered the arm and disengaged itself. (d) After giving impulse, the linkage has jolted the escape pin and so released the detent. (e) The escape pin is resetting the arm and the linkage pin is resetting the detent.

I was delighted with the way this remontoire operated once it had been fitted with the *fly* shown in the full drawing, Figure 12.4. Clockwork which has a steady plodding job of work to do needs no fly, because the load keeps the available power under control. The task of a remontoire is more like a sudden exertion, using an excess of power in a short interval, and in practice it always seems to demand some kind of moderator to control the action.

Without its fly, my remontoire goes into a state of oscillation. The gravity arm is thrown up by the escape wheel with such force that it cannot stop abruptly when it reaches the rest position, but continues to rise under its own inertia. This releases the ear from the escape pin and causes the arm to drop a second time. Assuming that the master pendulum is absent when this experiment is being conducted, the remontoire will be activated again and again. I had vainly hoped that this would not happen, as I had arranged the geometry of the linkage to reduce the rate of lift as the resetting neared completion. Indeed, if the escape wheel is turned at a steady speed, the lifting comes virtually to a

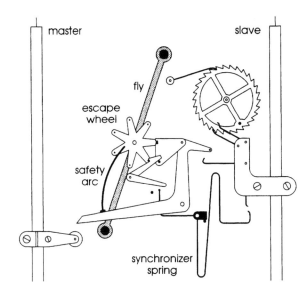

FIGURE 12.4
Full plan of W5.

standstill before the escape wheel is actually locked. However, the escape wheel does not in practice turn at a constant speed; naturally enough it accelerates.

I have often wondered whether it might have been preferable to design the linkage the other way round, with the lift biased towards the end of the reset rather than its start. The escape wheel would then accelerate from rest but would be restrained by an increasing load as the resetting progressed. This idea was rejected because it opened up the possibility of a reset starting but failing to run to completion, which would be inviting disaster.

The inertial fly is a double arm with weights on each end, mounted friction tight on the escape arbor by means of a coil spring and fibre washer. During the reset, the drive is not so forceful as to overcome the friction, so the fly's inertia can and does moderate the acceleration, which prevents the gravity arm overshooting. When the reset is complete, the friction grip allows the fly to slip on its arbor, so limiting the shock as the escape pin impacts upon the face of the detent.

At the experimental stage, one fiddles with the escapement enough to tempt providence, and it is easy to find the train running away. This can happen in the best regulated circles – even with Harrison's grasshopper escapement – and it is dangerous. A pin wheel turning at high speed can do considerable damage to any delicate parts that get in the way, and will almost certainly destroy itself in the process. In panic one tries to seize some vital part of the mechanism all too late. To prevent such disasters I found it worthwhile to design the *safety arc* to be seen on the drawing. In appearance, it looks like some kind of detent – as indeed it is – but in normal going, it never comes into play.

The very last part to be made was the hit-or-miss synchronizer, which is nothing more than a pair of bent wires rooted in a block of metal on a pivoted arbor. When the detent

The pendulum rods, winding barrel and double gravity escapement of W5 are set in front of the clock plates so as to be visible within the chapter ring. The Harrison maintaining power, wheel train and fly are hidden from view behind the main clock plate.

drops, it strikes the tail end of the synchronizer, which tips up the other end and causes it to catch – or fail to catch – on a tab protruding from the bracket on the slave pendulum rod. Some relaxing research went into the shaping of the synchronizer spring so that it would extend itself in a dignified manner. The part most easily seen from the front of the clock happens to be the upper U-bend, which looks remarkably like a common paper clip.

There is more to designing a synchronizer than shaping a wire; the vital thing is the phasing. Ideally, the synchronizer should come into operation as the slave pendulum passes through its centre of swing, so as to increase the restoring force throughout a full excursion of the pendulum from centre to extreme of arc and back again, where the hook lets go. By a happy accident, the pendulum frequencies of 1 hertz for the slave and 0.7 hertz for the master, suited the phasing of events to a tee, Figure 12.5. The fact that the gravity arm first

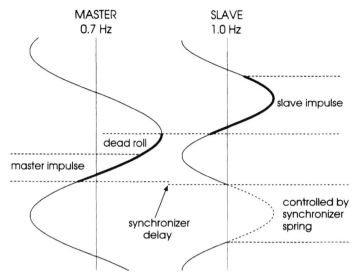

FIGURE 12.5
Phase diagram for master and slave pendulums of W5. (Displacement is plotted horizontally and time vertically.)

contacts the master's impulse roller at the extreme of its swing can be criticized for giving an excessive amount of dead roll, but this solves a question that can too easily escape attention. What is the routine for starting up a synchronized master and slave system?

The two pendulums must be started up in something like their correct relative phasing. For W5 the rule is to hold the master pendulum at its extreme of swing until the slave lowers the gravity arm. As soon as the arm touches the impulse roller, the free pendulum is released. The phasing is now close enough for the synchronizer to draw the pendulums into accurate synchronism over the course of the following ten minutes or so.

Even without engineering drawings or detailed specifications, sufficient information has now been given for an enthusiastic constructor to make another version of W5.

With a master pendulum weighing only one kilogram, it is hardly to be expected that W5 would match the performance of high-grade regulators having pendulums many times heavier, but freedom does count for something. The unloaded Q of the heavy seconds pendulum described in chapter 2 was 14 000, but after loading it with the task of driving a count wheel, it fell to 8400. This is 60% of the unloaded Q, a percentage serving as a measure of the pendulum's freedom. The unloaded Q of W5's lighter pendulum was only 10 500, but the loading reduces it to 9000, giving 85% as the figure for freedom. It may be of interest that the unloaded Q of the *slave* pendulum in W5 is 10 000, which drops to 4250 when loaded. The freedom figure for the slave is therefore 42.5%, or half that of the master. It would be interesting to know the figure for a Synchronome; I estimate that for the Shortt pendulum as about 97%.

A refinement which I consider worthwhile for any precision pendulum open to the atmosphere is a barometric compensator possibly employing aneroid capsules. These are sealed metal containers, each about the size of a tea biscuit containing little more than a vacuum, sprung outwards to prevent implosion from the pressure of the atmosphere. With a diameter of about two inches, each circular face is subject to a force of about 50 lb weight from the atmosphere, and naturally enough the thickness of the capsule varies as this force varies. If a capsule were totally evacuated, its thickness would vary also with temperature, because of the varying compliance of the metal, but this undesirable side-effect is avoided (as I understand it) by leaving a little air inside. The pressure of the trapped gas increases with temperature, compensating for the extra compliance of the metal.

To compensate a pendulum, it is usual to mount a pile of capsules, topped by a suitably chosen weight, on a bracket somewhere near the top of the pendulum rod, Figure 12.6. When the barometric pressure is low, which would normally cause the clock to gain as explained in chapter 2, the aneroids expand, the weight rises slightly and lengthens the period of the pendulum. Many people find it surprising that the raising of the weight causes a loss. They would argue that it raises the centre of gravity of the pendulum as a whole, thereby effectively shortening and hastening it. The rising weight does indeed raise the centre of gravity, but if placed near the top of the rod, it *lowers the centre of oscillation* and has a retarding effect. The point to remember is that the centre of oscillation, which determines the period, depends both on the centre of gravity of the pendulum and upon its radius of gyration (see Appendix).

Let us not go too deeply into the mathematics here. Rawlings shows us that the *addition* of a small weight (for the purpose of regulation) at some point between the suspension and the bob always makes the pendulum go faster, and has its maximum effect when attached half-way down the rod. In a barometric compensator the extra weight is there the whole time, but it is a moving weight. Because the effect of its presence is greatest at the half way point, any movement of it *upwards or downwards* from that point will cause a relative loss. The effect is more and more pronounced the farther the weight is from the centre, continuing indefinitely in both directions, even above the point of suspension or below

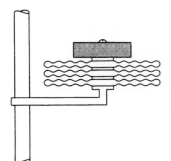

FIGURE 12.6
A pile of aneroids, topped by a suitable weight and fixed near the top of the pendulum rod, acts as a barometric compensator.

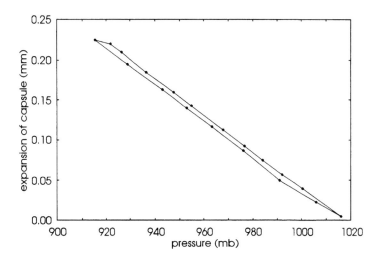

FIGURE 12.7 *Calibration of an aneroid capsule.*

the bob. Note that a compensator positioned below the centre of the rod must be upside down compared with the one in the diagram.

The compensator for my W4/W5 pendulum had its origins in a catalogue of goods and materials surplus to the requirements of World War II, where I spotted an advertisement for a packet of aneroid capsules costing no more than a few pence. I knew in my bones that aneroids would be difficult to procure if ever I should need them, so I ordered a packet. When it arrived, I examined the capsules with interest; they looked flawless and of high quality, smeared in grease and wrapped in layers of oiled silk as though destined for some tropical outpost. The label giving 'coefficient 0.054' in no stated units meant nothing to me, but I was fortunate in having a contact in the aerospace business. Obligingly he popped one of my capsules into a pressure chamber and calibrated it, Figure 12.7. This was precisely what I needed, and even showed the hysteresis, i.e. what goes up does not immediately come down, nor vice versa! The expansion coefficient turned out to be 0.002 25 mm of thickness per millibar reduction in external pressure. By one of those coincidences which delight numerical analysts, this happens to be 0.003 000 inches of thickness per inch of mercury, but no amount of playing with units has yielded the coefficient 0.054 quoted on the label.

While my old W4 had been working after its fashion, I had been able to measure its barometric error as −0.012 s/day per millibar. Armed with this and knowing also the coefficient for the capsules, I had all the data needed to design a compensator. The fact that the diameter of the capsules tallied with that of my pendulum bob suggested a neat arrangement with the compensator fitted underneath the bob, Figure 12.8. A single capsule with a modest weight on its under surface was sufficient for my relatively light pendulum. However, there was no certainty that the same pendulum in the new clock would have the same barometric error, and in fact it turned out to be about 25% smaller. The

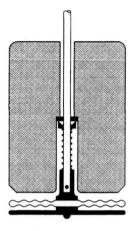

FIGURE 12.8
Section through the bob of W5. Its three-point support is represented here with artistic licence.

flotation and inertia components touched on in chapter 2 would of course be the same, but with W4's recoil escapement any changes of amplitude caused by pressure variations would have caused significant variations of escapement error. In W5, this component of barometric error was entirely absent. After reducing the compensator weight to about 40 g, no barometric error has ever been detectable in W5.

My principal aim in designing this pendulum was to avoid any stick–slip effects between the brass bob and the invar rod resulting from differential expansion. To this end, the bob touches the rod at one narrow neck only, at the centre of the bob. By a lucky accident for which I have no explanation the pendulum requires no temperature compensation. Compensating tubes of different lengths were tried, but better results were obtained without them.

As may be seen, the bob is bored with the traditional shoulder in the centre. A glass-hard steel collar resembling a washer is fastened to the shoulder with epoxy resin. The hole in this 'washer' is a sliding fit on the rod, and is the only area of contact between the bob and the actual rod. The washer rests on a hard steel collar (not invar) glued to the pendulum rod with anaerobic adhesive. This has a raised edge, filed down before hardening except at three points equally spaced round the circle, providing the bob with an inverted tripod on which to rest. The purpose is to prevent the bob from rocking, even by a microscopic amount, as the pendulum swings. I subsequently discovered that a similar possibility had been spotted by Sir George Airy, who wrote a little-known paper in 1829 'on a correction requisite to be applied to the length of a pendulum consisting of a ball suspended by a fine wire' (Airy 1829).

Under the collar a stiff coil spring serves two purposes. If the collar were not glued to the rod, the spring would be of such a strength as to support the bob, which means there is no shearing force on the glue, and hence no tendency for the collar to creep down the rod as a result of repeated differential expansion and contraction. Anaerobic adhesive is highly effective, and does not really call for any such safeguard, but there it is. More significantly, the spring exerts a force of 1 kg on the sleeve pinned to the bottom of the rod, into which the aneroid capsule is screwed. There is no glue here, and the spring takes up any slight play there may be in this fixture, once again to prevent swaying or rocking.

While bringing the pendulum to time, an extra sleeve replaces the spring, and the collar rests loosely on it. The clock is then tried, and the sleeve gradually reduced until the period is about right. Glue is then applied to the collar but not to the sleeve. When the glue has set, the sleeve is removed and replaced by the spring. This operation does of course alter the frequency very slightly, but final tuning is always carried out by means of small weights on a regulating tray.

Although it has never been publicly exhibited, W5 gained a little notoriety by winning the first of a series of international competitions organized by the British Horological Institute for 'Horology of Outstanding Interest'. One of the consequences was to bring me into contact with more horologists than I had ever known before, both professional and amateur.

As for timekeeping, W5 remains something of an enigma. At quite an early stage it showed its prowess in a remarkable period of about 150 days selected from a longer run, during which it gained at an astonishingly steady rate, the indicated time straying from absolute linearity by less than half a second either way over the whole of that period. The graph in Figure 12.9 covering that period is *not a rate chart*; it is the observed departure of the clock's time from a straight line graph of error. The straight portion can be seen in the latter part of the full run in Figure 12.10. It should be clear from this larger context that the rate of W5 can vary by as much as a minute a year in the long term, which points up the distinction between short-term and long-term stability. Other stretches showing exceedingly good short-term stability have been recorded, such as the early part of the run shown in Figure 14.2, which again stays within ±0.5s of a straight line for 150 days, but crucially no such performance can yet be guaranteed in advance.

Several far-fetched hypotheses have been suggested for lapses in the behaviour of W5 – instability of materials, gravitational anomalies in the vicinity of the Malvern Hills, interaction of a magnetized pendulum rod with fluctuations of the Earth's magnetic field, and

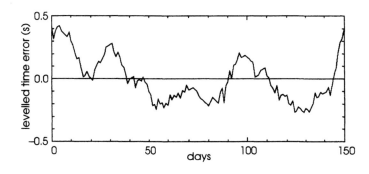

FIGURE 12.9
Residual error of W5 after allowance for rate.

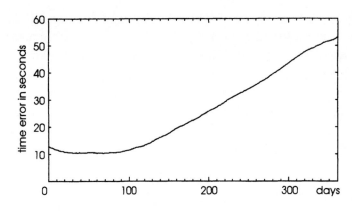

FIGURE 12.10
A full run of W5, 11-8-86 to 6-8-87, from which Figure 12.9 was selected.

earthquakes. All such excuses will have a familiar ring to other pendulum buffs, but the true explanation may lie closer to home.

The importance of a solid support for a precision pendulum has been emphasized by Rawings (1993, p. 40), and I have myself analysed some consequences of the lack of it (ibid., p. 100). However, as a reaction against the massive pendulums and bolted brackets of the traditional regulator, I had wished to see what could be achieved more simply. Both pendulums of W5 are supported by the clock casing, whose back simply hooks on to a pair of screws in the wall. To eliminate wobble, pads projecting from the back of the case are thrust against the wall by screws in the case, tightened from the front. Simple and convenient as this is, it is by no means ideal for a precision pendulum clock and may yet prove to be the dominant factor limiting the performance of W5.

CHAPTER THIRTEEN

Error correction

No sooner have the hands of a watch or clock been set to the correct time than they will start to wander away from it, and once they have lost the place, so to speak, there is no magic by which they can find it again other than by accident. This is an important point of principle. An independent timepiece is constantly predicting the time by a process of dead reckoning, and it is not to be wondered at that the longer it goes without checking, the farther from the truth it will be. Much of the interest of error analysis stems from the fact that different sources of error grow in different ways and at different rates.

After an appreciable error has accumulated, we must choose between three courses of action: to reset the hands, to re-tune the resonator, or to leave everything alone. Those who look after Big Ben keep adjusting the resonator by altering the regulating weights on the pendulum, which must be easier than resetting the clock by a couple of seconds (the maximum error tolerated at the strike). However, Big Ben is not the kind of clock that occupies much of my attention. For a proper study of errors, it is advisable to leave the resonator strictly alone. It is also a good idea to leave the hands alone, as this will simplify record keeping. After a while, a daily gaining or losing rate becomes clear, for even in the absence of every imaginable defect, the underlying rate of a clock can never be zero, no regulation ever being plumb perfect.

Let us begin with just two sources of error, an offset at the start of a trial, and a daily gaining or losing rate. Over a long enough period the rate will always win, even though the hands are initially offset in an attempt to allow for it, as shown for example in Table 13.1. After a week, the effect of a losing rate is here the predominant cause of error.

Some of the early Shortt clocks suffered from an even worse peculiarity called drift, which is a steadily changing rate as though from a shrinking or stretching pendulum rod. The latter was indeed the explanation offered by the Synchronome Company for the deceleration exhibited by at least three of the early Shortt clocks, SH3, SH4, SH11 and

Error (s)	3	2	1	0	−1	−2	−3	−4
Rate (s/day)	−1	−1	−1	−1	−1	−1	−1	−1

Table 13.1 Predominance of rate

perhaps all those between. My own invar rods have shown evidence of *shrinkage* over a period of years, causing W5 to accelerate. This defect is more serious than any fixed rate because it cannot be remedied by any mechanical adjustment of the clock, and it will eventually swamp errors of the types already mentioned.

In the 1920's, Professor Sampson of Edinburgh had custody of two Shortt clocks, the first of which, SH0, was made by W. H. Shortt himself. Later he acquired SH4, which was one of those afflicted with a drifting rate (Sampson 1928). The drift seemed absolutely steady and could therefore be allowed for by calculation. In an observatory, such a thing is well worth doing, as the disentangled time might be more accurate than that from any other source but the stars. However, we are not all astronomers, and I was once sharply reminded by Martin Burgess that the general public simply wants the *time*, not data from which the time can be *calculated*. For my own clocks, I am quite happy to allow for a steady gaining or losing rate, provided it is not too large, but I do object to having to allow for drift.

As already remarked, a steady drift must eventually swamp any non-zero rate, however large. (A car continuously accelerating eventually overtakes any car being driven at a constant speed, however fast.) With a steady rate, error is linear in the sense that a chart of the error would be a straight line with a slope equal to the rate. The error grows in direct proportion to the time, but it might not be quite so obvious that the error caused by a drift of rate is quadratic, i.e. involves the square of time. To see this, assume a constant drift, as in the Table 13.2, construct a set of rates to tally with it, and finally a set of time errors.

This is a particularly simple example, but no matter what the starting error or the starting rate, the quadratic effect will eventually predominate. When charted graphically, a quadratic dependence on time shows up as a parabola. Conversely, any error chart of a parabolic shape indicates at once the presence of rate drift. The parabola may not start at its vertex as it does in Table 13.2, and may be either way up. If the error chart is bowl shaped the drift is positive (acceleration), if umbrella shaped, it is negative (deceleration).

Drift (s/day/day)		2		2		2	
Rate (s/day)	1		3		5		7
Time error (s)	0	1		4		9	16

Table 13.2 Predominance of drift

ERROR CORRECTION

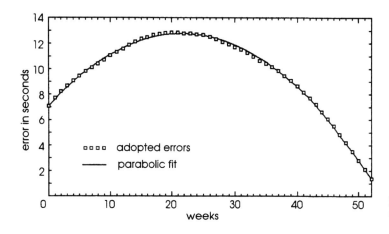

FIGURE 13.1
Time error of Shortt No. 4.

The errors adopted weekly for Professor Sampson's SH4 clock exhibit the presence of negative drift (deceleration) extraordinarily clearly, Figure 13.1. The continuous curve is the parabola which was calculated to fit the figures for error, adopted weekly, as closely as possible. The phrase *adopted error* is frequently encountered in astronomical literature. The purpose of a primary standard is to help in defining the time here and now, and when all the evidence has been gathered, the astronomer must try and decide what the time really is. The difference between the best estimate of the time and the reading obtained from the clock in question is the error 'adopted' for the clock. It is a nice point for the layman to consider!

I have taken a liberty with this historic record of SH4, but one to which Professor Sampson would not have made the slightest objection. In order to keep the clock's error within reasonable bounds, Sampson enhanced the vacuum in the clock case on two occasions in the year, thus abruptly altering the rate by changing the barometric error. Using his own figures, I have reconstructed the time error which would have been recorded had these changes not been made. The actual numbers written on the error axis of my chart are therefore quite arbitrary. In any case, Sampson's own published chart is marked off in seconds without any numbers against the divisions. They are irrelevant to the matter in hand; I have marked the vertical axis with arbitrary figures only to show that each division represents one second.

The actual amount of SH4's drift was only –0.5 milliseconds per day per day, but this produces a loss of 33.3 seconds after a year! One can see from this how potent an enemy of good timekeeping a drift of rate can be. The object of fitting the parabolic curve was, of course, to enable the systematic error to be subtracted out. The tiny remaining discrepancy, Figure 13.2, shows at once the remarkably high accuracy of SH4. After allowing for the drift, the clock was never more than a quarter of a second away from a straight line chart throughout the whole year.

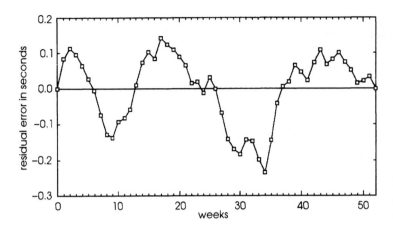

FIGURE 13.2
Residual error of Shortt No. 4.

The drift in Professor Sampson's Shortt clock was a spontaneous form of *systematic* error. Errors caused by external factors such as temperature fluctuations may also be classified as systematic, for even though the temperature itself may be unpredictable, the clock's response to it should be predictable once the coefficient has been established. The design of a compensator relies on just such an evaluation, which should always be the outcome of a practical experiment.

For a period of time, the clock's rate must be observed along with the temperature, and the extent to which the rate follows the temperature estimated. To give a sense of the accuracy needed, an ordinary steel pendulum has a temperature coefficient causing a change of rate of about -0.5 s/day/°C, but a clock with any pretensions to accuracy should be compensated to within about ± 0.01 s/day/°C. This figure is based on the specifications used by the famous German firm of Riefler, all of whose pendulums were guaranteed to within ± 0.02 s/day/°C, whilst their best ones were within ± 0.005 s/day/°C (Riefler 1981).

Experiments with heat are surprisingly difficult and cannot be hurried. As it is not practicable to place a pendulum clock in a refrigerator or an oven, the variation of temperature between day and night might seem to be a useful starting point, but I have never succeeded in turning this idea to good account. A longer spell of heat or cold works better, as though the clock needed more than a few hours to settle down. There is little choice but to take measurements daily and rely on spells of hot and cold weather to provide the necessary variety. Ideally, the experiment should start a few days before the onset of a heat wave and continue for a few days after its cessation, which presents a certain logical difficulty. In the hope of interesting or helping amateur clockmakers who are confronted with the problem of making do with an ordinary spell of varied weather, I shall describe in some detail the analysis of a fine set of observations made by James Chandler, a keen amateur horologist whose clock CM4 was under test in preparation for the improvements he planned to make to its temperature compensation.

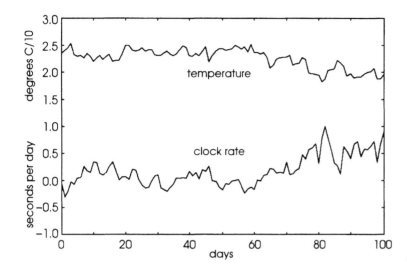

FIGURE 13.3
One hundred days of
James Chandler's 'CM4'.

The data for CM4 consisted of error readings, maximum and minimum temperatures, and estimates of the mean barometric pressure for each day. Rightly or wrongly I decided to ignore the barometric observations to begin with. My first step was to estimate the daily mean temperatures from the observed maxima and minima. The recipe I use for this – two thirds of the day-time maximum plus one third of the night-time minimum – is crudely based on the assumption that day temperatures persist for twice as long as those at night. To go with these, I would need the clock's daily rates, which are easily obtained as differences between successive error readings. The two sets of data are charted in Figure 13.3, where, for the convenience of using a common vertical scale, I have divided all the temperatures by ten. This relative scaling happens to make the fluctuations of rate numerically comparable to those of the temperature, indicating immediately that if the rate is indeed varying with temperature, the coefficient cannot greatly exceed ±0.1 s/day/°C. One might perhaps guess at ±0.15 s/day/°C, but until the detailed structure has been compared, it is impossible to say more.

To look for correlation between two sets of measurements, the usual procedure is to plot one against the other. For a hundred days' observations of rate and temperature, there will be a hundred points to plot, and in a perfect world they would all fall on a straight line whose gradient would, by definition, be the temperature coefficient for the clock. However, other disturbances are always present, and we inevitably finish up with a cloud of points, sometimes called a *correlogram*, Figure 13.4. The fact that the cloud is on a gradient shows that the rate varies with temperature. The straight line with a slope of −0.15 s/day/°C is quite a good fit.

Had we been plotting the heights of randomly selected adults against their weights, there could be no objection to the use of this technique, but for the present purpose there

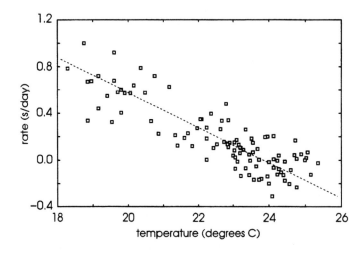

FIGURE 13.4
Chart showing statistical relation between rate and temperature.

is one possible drawback, in that the method throws away some of the data. What is lost is the time sequence of the observations, and even though I may choose to disregard it, I prefer it to remain visible throughout the investigation. On my computer screen, therefore, I always draw up two rate charts, one to show the variations in rate of the actual clock throughout the course of the trial and the other to show how a hypothetical clock would have behaved if its rate had depended solely on temperature. Different temperature coefficients can be tried for the imaginary clock, until the two charts most closely resemble each another.

Following this line of thought, the charts in Figure 13.3 can be regarded as a preliminary attempt, but the coefficient of +0.1 s/day/°C was clearly quite wrong. In Figure 13.5 I have remedied this by giving the imaginary clock the coefficient −0.15 s/day/°C. Although I took this figure from the correlogram, it could in principle have been arrived at by trial and error, repeatedly rescaling and replotting the 'temperature rates' until they best fitted the actual ones. By hand this would be a laborious business, but a computer (suitably programmed) makes light work of it and turns a chore into an entertainment! The two charts in Figure 13.5 have been brought to the same level by plotting only the fluctuations from the overall means, a process that equalizes the levels of the two charts in the most satisfactory way. Haphazard though the fine structure might appear to be, the quality of the overall fit is remarkably good, in spite of my having neglected barometric error.

The barometric coefficient for a pendulum clock is usually about −0.01 s/day/mb, but it can vary considerably from one clock to another. When two unknown coefficients have to be found from one set of data, they should be found simultaneously, not in sequence, as there may be some statistical relationship between them. It is, for example, quite possible that the higher temperatures will occur during periods of high barometric pressure, and if this is so, separate evaluations of either coefficient will surely give misleading answers. No

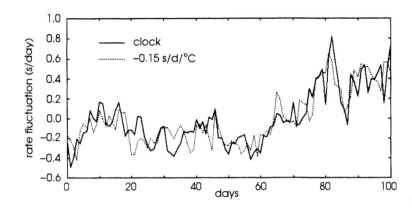

FIGURE 13.5
Plots of rate fluctuations and temperature fluctuations.

method of analysis can be expected to apportion error correctly to one cause whilst another closely related cause has been suppressed.

The problem of finding two or more unknown coefficients simultaneously is common enough in science, and a favoured method of solution is known as *least squares analysis*. To vary more than one coefficient at a time, graphical trial and error techniques are much too laborious, but any number can be varied algebraically by means of the calculus, which is how least squares analysis works. Important though the technique can be for those striving to attain high accuracy with pendulums, it never figures in books on horology, so the following brief description may be found useful.

The mean values of rate, temperature, and pressure are of no interest and must first be subtracted out, leaving three sequences of fluctuations which I shall denote by R, T and P respectively. The hypothesis is that T and P contribute to R in certain proportions which I shall denote by x and y. These are the unknown coefficients for T and P, and if there were no other sources of variation in R whatsoever, the following equation would hold good on each day:

$$R = xT + yP.$$

To solve for the two unknowns x and y, only two such equations are needed, but with a hundred days to play with, we are spoiled for choice. In theory, we could pick any two days we liked, solve the equations, and always reach the same answers. That is not the way of things, because the rates will surely have been influenced by other factors. Indeed, we cannot hope to satisfy all or even three of the equations exactly, so we must be content with making the residual rates,

$$R - xT - yP$$

as small as possible overall. In a run of a hundred days, there are a hundred of these residuals. Least squares analysis is so called because it minimizes the sum of their squares. Commonsense supports this idea, because it scores positive and negative residuals equally, and prevents them from cancelling each other out. (The plain sum of the residuals is zero when the mean values of R, T and P have been subtracted out.) This completes the statement of the method.

There is a sound basis of statistical theory in support of least squares analysis when the residuals are due solely to random experimental errors. It is less easy to justify the method if the residuals are due to the neglect of some systematic error, though it must be admitted that most experimental physicists are happy enough to use the method without asking too many questions. The actual mathematics summarized in the following short paragraph can be skipped by those readers who have come so far but baulk at going much further!

Using the *sigma* notation, the sum of the squares of the residuals is $\Sigma(R - xT - yP)^2$. To minimize the sum, we differentiate it with respect to x and to y, and equate the resulting expressions to zero, which results in the following pair of linear simultaneous equations

$$x\Sigma T^2 + y\Sigma PT = \Sigma RT$$

$$x\Sigma TP + y\Sigma P^2 = \Sigma RP.$$

These equations can be solved by the usual method taught at school; the main labour is in the preliminary work of evaluating the sums of products. These are taken over all the days of the run, and so use all of the available data. Had there been three unknowns to find, there would have been three equations of a similar pattern. When there is only a single unknown – say the temperature coefficient – the single equation is $x\Sigma T^2 = \Sigma RT$, and the solution can of course be written down immediately.

In passing, it should be mentioned that this last formula gives the temperature coefficient of CM4 as –0.13 s/day/°C, which is not far from the value estimated graphically. However, when least squares analysis was applied to both coefficients, the barometric rate came to –0.014 s/day/mb and the temperature rate fell to –0.096 s/day/°C. This is quite a significant change; to understand how it comes about, one has only to chart the temperature and barometric contributions separately, Figure 13.6. The similarity between these charts explains everything. When the barometer was ignored, temperature was being asked to shoulder the combined variation, and it succeeded in doing so remarkably well.

Ideally, one would like to present within a single frame a superposition of the clock's own rate fluctuations, the fit obtained from the appropriate combination of temperature and pressure, *and* the separate contributions of each. However, when four such charts such as these are superimposed, it is impossible to disentangle them by eye, even in colour on a computer screen. This came home to me most forcibly when I analysed the rate fluc-

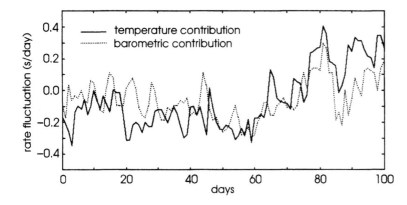

FIGURE 13.6
Contributions of temperature and pressure variations to rate fluctuation.

tuations of a regulator clock by William Hardy now belonging to the National Museum of Scotland. This had been carefully observed for over fifty days whilst under restoration by John Redfern, who unusually and painstakingly made observations of error, temperature, pressure, and – most remarkably – arc of swing to a precision of a thousandth of a degree! There were three coefficients to be found, and superposition of all the resulting rate charts would have merited a prize at the Tate Gallery (Woodward 1993).

Happily, there is an easy way around this difficulty. By forming the cumulative sum of the rate fluctuations – not the actual rates – a quasi-error chart is reconstructed. This differs from the actual error chart only in having had the mean rate subtracted out. In other words, the error chart has been levelled out; there is no overall gain or loss between the start and end of the run. If the first error is chosen for convenience to be zero, the last error will be zero also. Such a levelled chart for CM4 is shown in Figure 13.7, along with

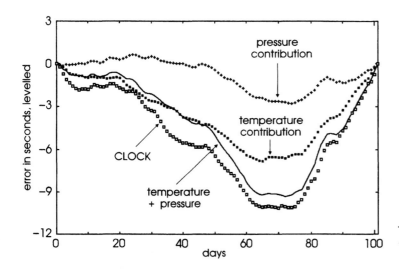

FIGURE 13.7
Reconstructed time errors and estimated contributions from temperature and pressure variations over a hundred days.

the similarly levelled fit obtained from the combination of temperature and pressure, whose separate contributions are also shown.

I consider the fit between the error of James Chandler's clock and the error resulting from temperature and pressure variations remarkably good, demonstrating the inherently high accuracy of the clock (whose compensation has since been improved as a direct result of this experiment). When comparing error charts such as this, one must not be unduly concerned if the curves run parallel to one another instead of coinciding. While running parallel, the rates correspond, so an offset is not in itself of any consequence. Here it was due to the fact that the rates did not correspond too well during the first three or four days of the run. We can also see how similar in shape were the separate contributions of temperature and pressure and we can judge for ourselves that a different choice of coefficients might have resulted in a fit almost as good as the mathematical optimum. However, that optimum turned out to be remarkably accurate, as much the same barometric coefficient was obtained from a later experiment.

Ideally, of course, temperature should be held constant during a pressure experiment, and pressure held constant during a temperature experiment, conditions not easily brought about in the home. My own solution has been to keep an eye on the barograph during the depths of winter, when the house can be kept at repetitive temperatures day and night by central heating. It is then a straightforward matter to determine the clock's barometric coefficient, after which I fit the pendulum with an aneroid compensator and cease concerning myself with pressure. Later, during the summer months, the temperature coefficient alone can be found if and when a heat wave should occur.

Being sceptical by nature, I doubt whether many clocks are as accurately compensated for temperature variations as they might be. It is a time consuming task, and progressive alterations to a pendulum can hardly be described as convenient. Lord Grimthorpe provided one solution to this in his book on *Clocks, watches and bells* in the form of an auxiliary compensator consisting of a bimetal rod above the bob, with sliding weights on the arms, Figure 13.8. The bimetal is so arranged that it bends upwards in heat, thus lifting the centre of oscillation of the pendulum and introducing a gain of rate, adjustable by altering the positions of the weights. This ungainly arrangement can be much improved simply by hiding it at the top of the pendulum, where it will be equally effective if turned upside down so that the arms droop in the heat. That the lowering of a weight fixed above the mid-point of the rod should quicken the pendulum has already been remarked upon in connection with barometric compensation in the previous chapter.

Grimthorpe rightly points out that it may be asking too much for a bimetallic compensator to deal with the entire temperature coefficient of a heavy pendulum on an ordinary iron rod, but it is well suited as an auxiliary compensator for an invar rod in which some residual error remains. For a light pendulum, such as that in my gearless clock (chapter 7), I have used the system as the sole means of compensation. (Suitably sensitive bimetal strip material is manufactured by Texas Instruments under the code TRUFLEX® P675R-TM2.) A neat extension of this idea was suggested to me by Henry Marcoolyn,

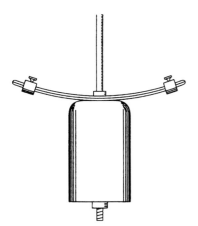

FIGURE 13.8
Grimthorpe's compound bar temperature compensator.

who proposed placing a tray for weights at the end of a bimetal strip. In this way, the temperature compensation can be varied without stopping the clock, and if the ordinary regulating tray is placed nearby, weights transferred from one to the other would alter the temperature coefficient without altering the overall rate.

CHAPTER FOURTEEN

Noise modulation

Speak to an audio engineer of 'white noise' and he will imagine a rushing sound not unlike what we imagine we hear when we put a seashell to our ear. By analogy with the colours that go to make up white light, the adjective 'white' means that the intensity or *power* of the noise is spread uniformly across the whole band of frequencies with which we happen to be concerned. In the correct jargon, the mean power density with respect to frequency is constant. On an oscilloscope, white noise from whatever source looks something like the record in Figure 14.1, but this particular trace shows noise over a band vastly lower down the scale than any musical sounds, lower even than the sounds of thunder or earthquakes. It is a plot of the rate fluctuations of Shortt clock No. 13 in 1938, compared daily with the 'rhythmic time signal' from the Rugby radio transmitter.* The frequency components to be found here are not to be measured in hundreds or thousands of hertz but in millionths! One millionth of a hertz is one cycle in eleven or twelve days.

It may seem strange to talk of frequencies so much lower than the fundamental frequency of the clock's resonator, but we are not really concerned with the actual frequency of vibration of the clock's pendulum, only with its variation from day to day, week to week, and even from year to year, as reflected in what horologists call its 'rate'. The variation can be analysed into components at numerous different frequencies in the range one cycle per day to one cycle per year. Coming to a topic such as this afresh, one can be muddled by the thought of discussing the frequency components of what is basically a

---

* The rhythmic time signal, intended for observations on the vernier principle, consisted of 61 pips per minute from a modified Synchronome clock controlled by a Shortt free pendulum, probably SH16 or SH49 (Humphry Smith, personal communication).

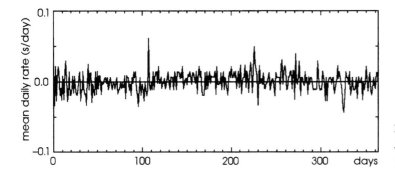

FIGURE 14.1
Daily rates of Shortt 13 (1-1-38 to 31-12-38).

variation in the clock's resonant frequency. When this happens I always find it helpful to think of an FM radio broadcast, where the radio frequency is made to fluctuate by the music. The sound of the oboe's tuning note of A at the start of a symphony concert comes over as a 440 hertz modulation of a radio frequency which might be centred on 100 000 000 hertz. The radio frequency then swings up and down 440 times a second. (How far it swings depends on the loudness of the note.) The applause at the end of the concert is a passable imitation of white noise, and the analogy with a clock is then close indeed. Give or take a factor of fifty million, the radio frequency is the frequency of the pendulum, and the audio modulation is the rate of the clock.

In the context of noise analysis, the words 'frequency' and 'rate' do not mean quite the same thing. If the frequency of a pendulum should increase by one part in a million, the clock's *rate of going* increases by just that fraction, from an indicated 1.000 000 seconds per second to an indicated 1.000 001 seconds per second, but the horologist's 'rate' is the 0.000 001 seconds of gain. This he would normally express as 0.0864 seconds per day, though it can be less confusing scientifically to call it one part per million (1.0 ppm). Rate is then the *fractional deviation* of the clock's frequency.

Random looking behaviour cannot properly be classified as noise unless there have been enough fluctuations to give it measurable statistical properties. The handclaps of an audience may be a passable form of noise, but a single handclap would be problematical. A good example of an isolated event in my own records occurred during the eighth timing run of W5, which is shown in Figure 14.2. This is a plot, not of rates but of actual time error, as also is the enlarged extract from it in Figure 14.3. It can be seen from the changes of slope that the rate suddenly changed, and about twelve days later, after the clock had lost precisely four seconds, changed back again. What are we to say about behaviour such as this? In Rawlings' memorable phrase,'clocks are kittle cattle', and the occurrence had to be classed as a freak event. It cannot be classed as noise if it never recurs. In noise theory, persistence is a concept of great importance.

The daily rates of SH13 present a tidy pattern of white noise which is most unusual for a pendulum clock. Those of SH41 shown in Figure 14.4 are puzzling for the opposite reason. This chart is nothing like white noise, but comes very close to what physicists call a

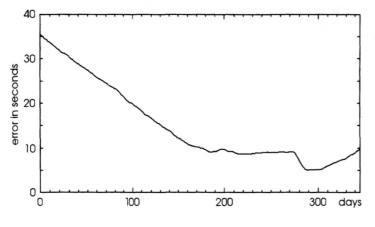

FIGURE 14.2
*Error chart for W5
(11-12-88 to 21-11-89)*.

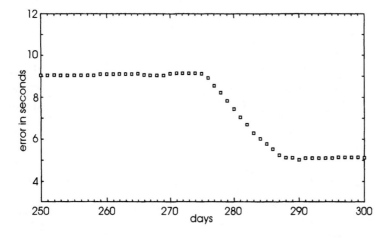

FIGURE 14.3
Detail from Figure 14.2.

random walk, a term which is almost self-explanatory. Each of these forms of noise can be best understood in terms of rules for their construction. A run of white noise can be simulated by tossing a die repeatedly and, after subtracting out the average value of 3.5 each time, plotting the scores one after another. It is easy to imagine a record such as Figure 14.1 as having being constructed in such a manner. Each day's score is independent of that for the day before, and the pattern proceeds with monotonous regularity. For a random walk, the scores are allowed to accumulate and might lead upwards, downwards or anywhere, like investments. The scoreboard for white noise is cleared to zero for each day's plot. For a random walk, it can be set to zero at the start of the run, but never thereafter. As a matter of interest, random walk frequency modulation is known to audio engineers as 'red noise' because it is exceedingly rich in low frequencies.

That clock rates should ever perform a random walk may at first seem extraordinary. To lapse into unscientific language, we are tempted to think that a clock knows in its heart of

NOISE MODULATION

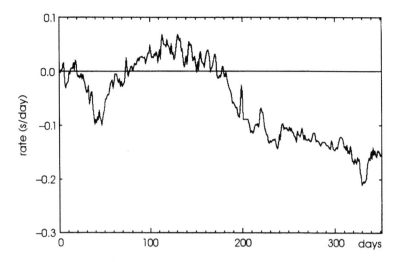

FIGURE 14.4
*Rate chart for SH41
(18-10-84 to 5-10-85).*

hearts what its 'true' rate is supposed to be. If its actual rate digresses a little one way or the other from the underlying true value, these are but temporary aberrations, as each day sees a fresh start. Unfortunately this is all wishful thinking, as few clocks are as well behaved as that. Atomic clocks can do it for remarkably long periods (relative to their periods of oscillation), but it is most unusual for pendulums.

This is perhaps the psychological moment to remind ourselves of the importance of distinguishing between a clock's *rate* and its indicated *time* error. The latter is really an accumulation of all its past rates, and if these had been modulated by white noise, the *time* error of the clock would be a random walk. To this we cannot have the slightest objection, for no clock keeping independent time can set its hands right spontaneously without referring to a time standard. If it did so, then – like a radio-controlled watch – it would cease to be an independent timepiece.

Returning to rates, there is something unbelievable about a clock's *rate* performing a random walk. Can a pendulum spontaneously stretch or shrink, and having done so, show no tendency to revert to its previous length? The rate of Professor Sampson's SH4 drifted, but the drift was steady, which made it predictable and therefore to some degree excusable. That *random* changes of length can also be irreversible seems much harder to accept, yet SH41 must indeed have been suffering from irreversible changes, not necessarily in the length of its invar rod, but in some one or more parameters affecting its rate. In this, SH41 was certainly not alone. An interesting consequence is that it does not make sense to try and determine the long-term rate of such a clock: the best estimate of its rate is always the one it has at the moment. Nothing from its past history helps at all, whereas the true rate for a 'white clock' is its average over the infinite past.

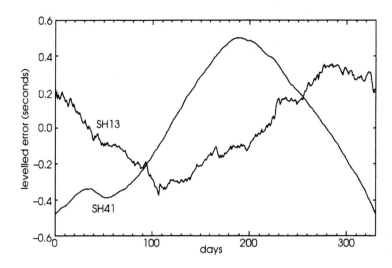

FIGURE 14.5
Error charts for two Shortt clocks.

It is interesting to speculate on why SH13 and SH41 should have been so different. Figures 14.1 and 14.4 do indeed look very different, but it is instructive to convert them both into records of cumulative time error and superimpose them, as in Figure 14.5. Apart from the small scale roughness of SH13, the two now look remarkably similar. As the two clocks were of identical design, one is strongly tempted to question whether the short-term roughness of SH13 was inherent in the clock or due in some way to errors in the time signals from the Rugby transmitter used as the reference. SH41 was measured with respect to an atomic time standard (Boucheron 1986, 1987), which puts this record of a Shortt clock in a class by itself.

Visual impressions are all very well, but in scientific work we need figures. Harmonic analysis to determine the colour of the noise may spring to the mind of some readers, but it is not the easiest thing to do, and it condenses the data hardly at all. What is needed is some way of jettisoning all the individual wiggles and abstracting the statistical structure. An extremely simple and effective way of doing this was originated at the US National Bureau of Standards (Allan et al. 1974a), and I shall always be grateful to Douglas Bateman for drawing my attention to the comprehensive NBS Monograph on the subject. The principle can be explained quite briefly, if not simply.

One starts from a sequence of daily observations and converts them into mean daily rates by differencing in the usual way. Each rate is then subtracted from that for the following day, giving a set of rate differences representing all the day-to-day rate variations. Here it may be well to interrupt the recipe with a comment. Those with a little knowledge of statistics may already be wondering why we should take differences between *two* of the observed rates, *both* of which will have been subject to a random variation. In elementary statistics, it is customary first to find the overall mean of a set of random numbers and use that as the base for measuring variations. Unfortunately this does not always work for clocks, for the reason already given – namely that many clocks cannot be said to possess any true long-term mean rate at all! It was to overcome this difficulty that Allan took rate *differences*, and made appropriate allowance for it at a later stage.

To continue, we have found the differences between each day's mean rate and that of the day following. All of these differences we square and add up, then divide by their number to obtain an average. This we halve in order to allow for the double random contribution included in each term. The square root of this figure is the notional uncertainty in the time error accumulated in a typical sample run of one day, a figure which comes to 0.000 77 s for SH41, as against 0.0135 s for SH13, a very big difference indeed. In scientific work, a dimensionless fraction is preferred, to obtain which we divide by the sample time, here 86 400 s. The resulting figure is the *fractional instability* for the chosen sample time. For SH41 it comes to 0.009 ppm for a sample time of a day, and for SH13 the figure is 0.16 ppm.

This calculation must now be repeated for other sample times because instability in the short run can differ markedly from that in the long run, and here I make no apology for repeating the recipe. To find the fractional instability for a sample time of a week, the clock's gain in a week is subtracted from its gain in the following week and as many such differences collected as are available in the data. These are duly squared, added up and divided by the number of terms. The result is halved, and the square root is taken. This is the notional uncertainty of the error accumulated in a week. To express it as a dimensionless fraction, we divide by the number of seconds in a week, the result of which is the fractional instability for a sample time of a week.

When this calculation has been repeated for a number of different sample times, a chart can be drawn up by plotting instability against sample time. For ease of interpretation, it is customary to use equal logarithmic scales on both axes. Obviously the sample time must always be considerably shorter than the total run of the data, so that enough differences can be collected to give a reasonably reliable average, my private rule being to allow for at least five non-overlapping samples in the run as a whole. This is why my charts in Figure 14.6 for SH13 and SH41 stop off at seventy days.

Slopes on a stability chart reveal the type of noise modulation affecting the frequency of the clock. Consider, for example, SH13 whose chart descends with a gradient of about one in two. Because of the equal logarithmic scales on the two axes, lengthening the sample time by a factor of 10 reduces the instability by a factor of $\sqrt{10}$. This is precisely the behaviour we expect from a completely random process such as white noise. (The fractional error when a randomly chosen sample of the population is polled is inversely proportional to the square root of the number of people in the sample.) If a stability chart shows a *positive* gradient of one in two, we are faced with random walk modulation. In this case, when the sample time is increased by a factor of ten, the instability *increases* by a factor of $\sqrt{10}$, which is strange indeed.

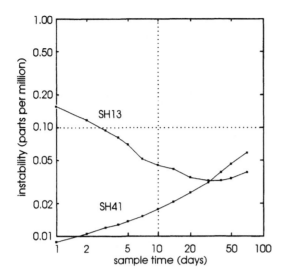

FIGURE 14.6
Stability charts for two Shortt clocks.

If instability seems a difficult concept to grasp when expressed in its fractional form, it is easy enough to convert it back into the notional accumulation of error over the sample time. All we have to do is to multiply the figure for instability by the sample time. This shows immediately that white noise frequency modulation produces an error increasing statistically as the square root of time, whilst random walk frequency modulation produces an error increasing as time to the power 3/2. This last is not quite as bad as the effect of a systematic drift of rate, shown in the previous chapter to give an error growing as the square of time.

In passing, it is useful to consider a downward gradient of one-in-one, or −45°. For such a slope, when the figure for instability is multiplied by the sample time, the answer comes out *independent* of the sample time. This can mean only one thing: it reveals errors of *observation*, not of the actual going of the clock. It is as though the hands were loose, an effect which can be caused in precision electronic oscillators by noise in the amplifier responsible for counting the vibrations. An example of such noise (Allan et al. 1974b) can be seen in Figure 14.7, where the product of sample time and rate instability is one millionth of a microsecond over three logarithmic cycles of the chart. With the apparatus used, the time simply could not be read any more accurately than that.

The bowl shape of this chart is not untypical. The −45° descent cannot persist for ever and eventually a flat section, known as the *flicker floor*, is reached. This quaint term derives from a type of noise modulation called *flicker noise*, whose properties are exactly intermediate between those of white noise and random walk modulation. Audio engineers call it pink noise. Flicker noise seems to be the norm for a high-quality pendulum clock and some examples of it are shown in Figure 14.8. Bateman's clock (Bateman 1994), being a precision clock open to the atmosphere, has been equipped with a barometric

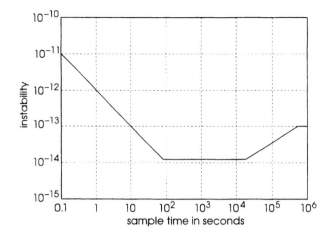

FIGURE 14.7
Example of a stability chart for a hydrogen maser oscillator.

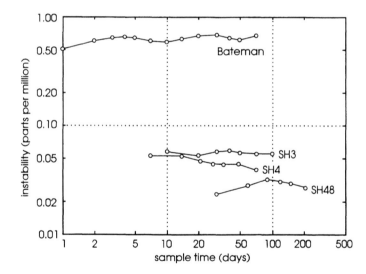

FIGURE 14.8
Examples of flat stability charts.

compensator, and has electronic amplitude control to avoid variations of circular error. The other clocks, being Shortts, were all *in vacuo*. Both SH3 and SH4 suffered from the systematic drift of rate mentioned in the previous chapter, but this was removed before the instability was calculated. The data for SH48 was taken from Rawlings, who had it from the US Naval Observatory, and to the best of my knowledge no corrections were applied.

A summary of the interpretation of gradients in a stability chart is given in Table 14.1.

Flicker noise is a widespread phenomenon by no means limited to clocks. It was first observed in the current passing through certain types of thermionic vacuum tube and has been aptly described as *noisy noise*. By a strange irony it is tantalizingly difficult to account

Gradient on stability chart		Type of modulation
−1.0	◹	errors of observation
−0.5	◹	rate modulated by white noise
0.0	level	rate modulated by flicker noise
+0.5	◸	rate performing random walk
+1.0	◸	rate steadily drifting

Table 14.1 Interpretation of gradients

for. Being intermediate between white noise and random walk, a clock whose rate is modulated by flicker noise does not clear the scoreboard each day, nor does it build on yesterday's score. It is extremely hard to imagine intuitively a half-way house between two such modes of behaviour! It becomes even more tantalizing when we discover that the statistical law for the growth of the error is not as time to the power 0.5 (white noise rate modulation), nor as time to the power 1.5 (random walk rate modulation), but simply in direct proportion to the time.

It used to be common for watchmakers to describe a high-quality watch as 'good to a couple of seconds a day', or perhaps they might say 'good to a minute a month'. The implication was that the error would be directly proportional to the lapse of time since the hands were set. A watch that has not been properly brought to time by regulation would behave in just that way, but I do not think this was what the watchmaker was meaning. I think he was referring to the random error, which he either assumed or knew from experience would grow in direct proportion to time. As soon as I knew a smattering of statistics, this puzzled me as it has puzzled many others. We thought that two seconds a day should build up to about eleven seconds a month $(2 \times \sqrt{30})$, because of the tendency of random numbers to cancel when added up. It was a case of a little knowledge being a dangerous thing, but it is still puzzling. How is it that random errors caused by flicker noise rate modulation can seemingly fail to cancel at all? And why should clocks and watches behave in a way that sounds so simple and yet is so obscure?

In trying to come to grips with this sticky problem, I have found the computer to be a great boon, though it took some while to discover a way of simulating flicker noise. This I shall describe in the next chapter. flicker noise is rich in low frequency components, for the power of the noise is distributed across the frequency spectrum in inverse proportion to the frequency f. Indeed, '$1/f$ noise' is the term often used for it. It is easy to see that this gives an equal distribution of power to each octave of the frequency scale, and here lies another conundrum, for the number of octaves below middle C is infinite! Where does it all lead? This is a question which has been asked over and over again, and it is not one to be answered in a sentence.

CHAPTER FIFTEEN

The enigma of flicker noise

Flicker noise presents applied scientists with a problem. It is a widely observed phenomenon by no means limited to clocks. Originally discovered in thermionic valves, it has turned up in many fields other than electronics – in physics generally, medicine, hydrography, the stock market, and even in modern music (a form of noise) – but it is strangely difficult to explain. A generic mechanism for it has been proposed (West and Shlesinger 1989, 1990), but through my own shortcomings, I have not yet succeeded in finding that explanation helpful.

Some investigators have sought to construct arbitrary mathematical models of flicker noise in the hope of elucidating a puzzling phenomenon. It so happened that I had such an idea myself, suggested by the abrupt changes of rate shown in Figures 14.2 and 14.3. From this starting point I was able to develop a model with the required $1/f$ spectrum of frequencies. Only in my subconscious was I aware that a similar basis had already been researched in a deeper and more abstract way many years previously (Mandelbrot 1967). There are numerous ways of modelling flicker noise, and the fact that some arbitrary mathematical process produces an end result with the correct statistical properties is of no especial importance unless it is simple and suggestive. The model I am about to describe is the simplest I have discovered, and it does what I wanted, which was to provide a fast algorithm for generating flicker noise in a computer. As is often the case, the structuring of the model taught me even more than did the actual running of the computer program.

Like so many models of flicker noise, this one relies on the principle of summing a series of independent noise processes. Some researchers seem to find this unsatisfactory on the grounds that anybody can model anything by starting with a box of bricks and a sufficient number of variable parameters. They would prefer an integrated model having no adjustments. I happen to believe, with West and Shlesinger, that flicker noise is in

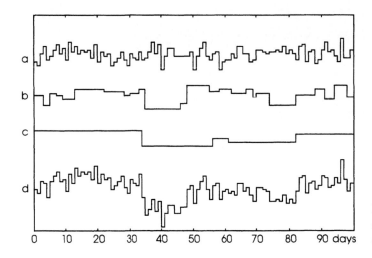

FIGURE 15.1
Stages in the synthesis of a model for flicker noise.

truth a composite phenomenon requiring a box of bricks, and my particular model has remarkably few adjustments. One could claim that it has none at all.

The first contributing noise process starts with a random number. A number thrown by a die would do, though it might be more acceptable to take the sum of, say, four dice thrown at once. The scores would then lie between 4 and 24, and would average 14. To avoid needless complications, I subtract out the 14 to leave a random number lying between –10 and +10 inclusive, with a mean value of zero. Provided the mean is zero, the actual distribution of possible scores is unimportant.

So far we have one random number. To proceed further with this first process, we must divide time into 'atomic' slices. In theory, these should be infinitesimally small, but the calculations would then take an infinitely long time! I shall take the indivisible slice to be one day. The dice are thrown each day, and the scores – in any convenient units – count as the first contribution to the required noise. The trace at Figure 15.1a shows the results obtained for this first process in a computer run simulating 100 days. It will be noticed that one day's noise may occasionally be exactly the same as that for the following day. This is because the chosen scoring mechanism always gives whole numbers, and there is a finite probability that we shall throw the same total twice running. With 'real' numbers rather than integers, such coincidences can never happen.

Matters must now be complicated by the introduction of what I visualize as illuminated signs saying 'WALK' or 'DON'T WALK', or, better still, *THROW* or *STICK*, one for each process. The sign must be consulted before any dice are thrown. Its mechanism operates randomly, but with an adjustable bias in favour of *THROW*. The first process had a bias of 1.0, ensuring a fresh throw every day. The second process, entirely independent of the first, has its own sign with bias set to 0.5, i.e. equally likely to show *THROW* or *STICK*. If the sign says *STICK*, the rule is to repeat yesterday's score. On average, therefore, dice for this second process will be thrown every two days. The third process has a bias of 0.25, giving fresh throws on average every four days, and so on. A run of this third process is shown in Figure 15.1b, and of the fifth at Figure 15.1c. Halving the bias for each successive process and starting with a bias of 1.0 shows that the adjustable bias is not so arbitrary after all; it could hardly be simpler. Finally all the processes are added together and the result, with qualifications yet to be described, proves to be flicker noise. The sum of only three selected processes, shown in Figure 15.1d, already begins to resemble flicker noise in appearance.

A run simulating 500 days is shown in Figure 15.2. I have reproduced the first process alone as a sample of white noise for comparison with the sum of thirteen processes beneath it. (Why thirteen is a question for later.) The look of white noise should now be familiar enough, but the second trace is new, and may interestingly be compared with the rates of Bateman's clock over an interval of 495 days, as shown in Figure 15.3. That this

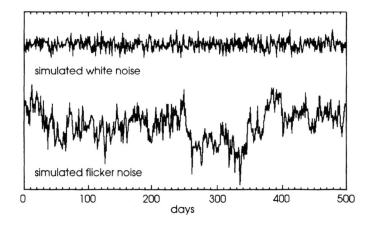

FIGURE 15.2
Simulated white noise and flicker noise.

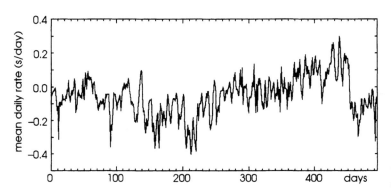

FIGURE 15.3
Flicker noise in practice: 495 daily rates of Bateman's clock over the period January 1974 to June 1975.

clock suffers mainly from flicker noise was already evident from the flat stability chart in Figure 14.8, which was computed from the same data.

Flicker noise simulated by the method described yields a reasonably flat stability chart when analysed in a computer, provided the runs are long enough to keep sampling errors to a low level. However, as a mathematician I prefer logical proof, and after several weeks of struggle I did complete such a proof. It showed the model to have a stability chart more or less flat from intervals of a day up to a limit that depends on the number of included processes. If that limit is removed, the chart is indeed asymptotically flat. It is slightly out of true over the first decade, and this was shown to be caused by the failure to subdivide time more finely than a day.

More conventional models of flicker noise are based on integrals over a continuous range of timescales, not on a sum of discrete processes at intervals of an octave. One might have expected the summation model to produce some form of ripple at octave intervals in the frequency domain or in a stability chart, but no such effect appears either in the full mathematical analysis or in any of the numerical runs. Why this should be is as yet an unsolved mathematical mystery.

I now return to the matter of the thirteen processes. This limit ignores high order processes that are unlikely to give rise to any *THROW*s at all, for when the bias has been halved enough times, the sign will always say *STICK*. For a run of 500 days, it is reasonable to limit the number of processes to thirteen because the fourteenth process would have a daily *THROW* probability of only 1/8192. The chance of a *THROW* in 500 shots can be worked out quite simply and comes to 6%. This result is e to the power $-500/8192$. The chance for the fifteenth process is 3% and so on. All these extra processes together increase the chance of just one extra *THROW* by 12%, which is negligible.

And now we must pause for thought. That we should discard all processes that will almost certainly fail to *THROW* during the course of the experiment may seem harmless enough, but each discarded process is stuck at some score between -10 and $+10$. Even allowing for the fact that these random scores tend to cancel, the sum of an infinite number of them will have randomly walked to infinity! The enormity of such an error seems more pardonable when we realize that it was perpetrated only once, on the first day, a day that is exceptional because the instruction to *STICK* makes no sense when there is no score to stick on. On the first day, therefore, every process must throw, and all we have done is to take a zero total score from the outset for all processes that are unlikely to change during the run. In the language of the electrician, we have discarded an infinitely large DC component from the spectrum of the noise. No model of flicker noise has any other choice; the discarded infinity may have been conceptually possible, but it is impossible in any physical experiment or computer simulation. We see now the wisdom of Allan's method of measuring instability: by taking the difference between the clock's behaviour in two successive time intervals, the awkward infinity is completely by-passed.

There are actually two infinities in flicker noise. The observed frequency spectrum of flicker noise is *1/f*, whose graph is a hyperbola, Figure 15.4. The area under the curve

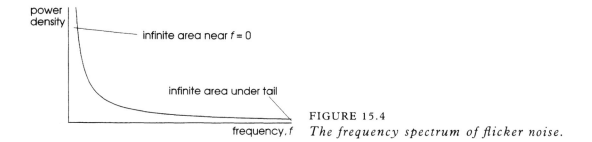

FIGURE 15.4 *The frequency spectrum of flicker noise.*

represents noise power distributed from $f = 0$ to $f =$ infinity. Awkwardly, the area is infinite at both ends! We have disposed of the low frequency end by neat if somewhat devious means, and must now consider the high frequency end. This is associated with short times which we have ignored by taking a day to be indivisible. Even in the basic physics of thermal noise this same dilemma occurs, for time can be sliced infinitely thinly and in thermodynamics each slice can be taken as a degree of freedom demanding its own ration of thermal energy. This was a dilemma of supreme importance in the history of science because it spelled the death of classical physics. The quantity of thermal energy per degree of freedom had to be reinterpreted in terms of discrete energy *quanta,* and the infinity then disappeared. In clockwork, of course, we are not concerned with thermal noise or quantum mechanics, and we can dispose of our infinitely short times on purely practical grounds.

My motive in designing a model of flicker noise was to find out how the random variations in the rate of a clock could result in a time error which – in a statistical sense – grows in direct proportion to time, seemingly with no benefit from cancellation of a positive rate on one day by a negative rate on another. That it does grow linearly is an experimental fact. We have only to look at the flat stability charts in Figure 14.8 to see it. As pointed out in the previous chapter, the notional error which accumulates in a time T is the product of T and the instability, which is calculated from observations taken over many such intervals. If the instability is independent of T, the error to be expected in a time T is proportional to T. Names like 'flicker noise' do not matter. The dilemma is staring us in the face.

As already remarked, cancellation of random numbers is a consequence of one number being opposite in sign to another; a clock's gain during one day can only be cancelled if it loses on another day. Any effect which tends to make a gain or a loss stick for a period of days will naturally reduce the amount of cancellation that will take place. The relevance of the model will now be clear. To take an extreme case, consider a high order process in which the chance of a *THROW* is minute. Day after day throughout the run this process is stuck on its initial score. Its contribution to the rate of the clock remains fixed, and its contribution to the accumulating time error grows in direct proportion to the time. Not all processes are as extreme as this, so it is still not obvious why evidence of cancellation is completely absent.

The key to this enigma can be found in a scheme which compares a run of a given length with one of twice that length. Comparing the nth process of the shorter run with the $(n + 1)$th process of the longer run, we find that the number of *THROW*s to be expected is the same for each, for in the longer run, the daily probability of a *THROW* has been halved whilst the number of opportunities has been doubled. However, the sum of all the daily scores for the longer run will be twice that for the shorter run simply because there are twice as many scores to add up. This is indeed the answer to the conundrum. The mathematician will see at once that a formal proof will involve problems of convergence, which it is beyond my power to express in ordinary words. At the high frequency extreme, there is the artificial limitation imposed by the time slice of a day (applying equally to all processes), which leaves one white noise process stranded without a partner. This is comparatively unimportant. The problem of divergence at the other extreme has already been circumvented by Allan's definition of instability.

Short of embarking on a full mathematical treatment, the matter must rest there. The model makes it quite clear that changes to a clock's rate that occur only very occasionally are the ones that cause errors to grow faster than they would if the rate turned over a new leaf every day. This conclusion is in interesting contrast with that of Rupert Gould (1960, p. 254), who writes as follows:

The best chronometer is that which changes its rate *least* and *slowest*. No machine has ever been made to keep exact time, and no machine ever had an unchanging rate. Its excellence as a timekeeper is determined by consideration of how much its rate alters, and how often. ...The deciding factor is 'rate of change of rate'.

One sees what he means, and I must leave it to the reader to decide whether there is any real incompatibility of view here. I do not believe it relevant that Gould was discussing marine chronometers with balance-and-hairspring movements rather than fixed clocks with pendulums. Three of the five or six chronometers I have analysed by Allan's method (Rawlings 1993, Fig. 18.5) show reasonably flat stability charts like pendulum clocks, and all can therefore be supposed to suffer primarily from flicker noise rate modulation.

To know one's enemy is the first move in the war against timekeeping errors. According to the late Professor Henry Wallman (personal communication), the prime objective for those who still conduct research in mechanical horology should be to eliminate flicker noise in favour of white noise, which gives ever improving stability as the length of a run increases. How this might be accomplished was never stated.

CHAPTER SIXTEEN

Wallman's conjecture

Clockwork may already have reached its apogee when it was eclipsed by electronics, but curiosity compels us to ask whether extra time might have improved the score. It is a question which has fascinated many scientific horologists. A generation back, the distinguished names which would have sprung to mind were those of Vannevar Bush and John Early Jackson in the United States, who were convinced that pendulum accuracy still had some way to go. Their well-known project was reported in *Scientific American* (Strong 1960), but as is so often the case the final outcome of the experiment is obscure; one can only presume that it proved a disappointment. It has been reported (Cain and Boucheron 1994) that Jackson was never able to adjust the spring in the ingenious compensator for circular error. Enthusiasts approaching the subject from every angle are too numerous to name, yet by present-day standards progress has been slow.

A precision pendulum swinging in air does not seem able to maintain a steady rate to closer than about half a part in a million, but the Shortt free pendulum swinging in a high vacuum was ten times more accurate. The usual scientific explanation for this difference is that accuracy depends on Q; in normal air we might achieve a Q of 30 000, but the Shortt pendulum had a Q of about 110 000. A high value of Q is certainly necessary, but is not sufficient in itself, for many other factors contribute to the stability of a pendulum clock. Systematic errors have to be eliminated or compensated for, and every parameter that goes towards determining the resonant frequency – principally the length of the pendulum rod – must of course be highly stable.

A high Q is often said to reduce undesirable interference from the escapement because it reduces the force needed to maintain oscillation, but there is another way of looking at it; the effect of a high Q is to set bounds on the natural tendency for the frequency of a resonator to wander. Let us start with the concept of a passive resonator in the presence of a noisy environment. At temperatures above absolute zero, every environment is noisy, for everything warm and capable of vibration vibrates the whole time. This applies to electrons, atoms, molecules, even pendulums, and in conditions of thermal equilibrium, each degree of freedom has the same average share of thermal energy. The ration is absurdly small for a large object such as a pendulum, and I do not suggest for a moment that thermal energy has any relevance to ordinary clockwork. It does however illustrate a principle. Whenever any kind of noise energy abounds in a system, it will tend to get shared out, setting up sympathetic vibrations in anything accessible. This is particularly noticeable in old cars, and pendulums are not immune.

Bateman (1993) has observed that the pendulum of his wall-mounted clock visibly swings through a tiny arc even when the power supply has long been turned off. The vibrations may originate from activities within the house, or perhaps from heavy road traffic. If the waveform were recorded and its frequency content analysed, it would no doubt be found to encompass the whole range of frequencies permitted by the pendulum's Q. With noise as its only stimulant, the frequency of a resonator can and does vary, wandering about aimlessly within a band whose width is about $1/Q$ of the mid-band frequency itself. I state this without proof, as it is a well-known result. With a Q of say 15 000 the frequency spread is equivalent to a variation in rate of several seconds a day — not bad, though easily bettered by the most humble of regulator clocks.

By driving a pendulum systematically, we try to keep its amplitude well above any noise level, and keep the frequency close to the centre of its band by taking the utmost care that impulsing should not disturb the phase. Indeed, the factor of improvement over noisy vibrations is determined solely by the accuracy of the phasing. That is what equation (10) in chapter 8 is all about; rough and ready figures put into that equation give an answer of just about the right order. Suppose, for example, that the escapement's impulse is timed correctly to about one degree of phase angle, or 1/360 of the pendulum's period of vibration. For a seconds pendulum (whose period is two seconds) this is about 5 milliseconds, which is 1% of the time taken for the pendulum to swing through the central 70% of its arc. It is a modest assumption, but when combined with a Q of say 15 000, the fractional time variation works out at about 0.6 ppm, or a twentieth of a second a day, which is about as good a result as is ever obtained by a pendulum in air.

The escapement or drive is thus the clock's saving grace, and everything would appear to depend upon it. If so, electromagnetic drive ought to score heavily over a mechanical escapement; using electronics it is surely possible to detect the phase of the pendulum and time the driving force to an accuracy much better than 5 milliseconds, yet strangely enough no spectacular breakthrough seems to have been achieved in this way. To the best of my knowledge, no pendulum swinging in the atmosphere has yet approached the performance of the mechanically impulsed Shortt clocks of the 1920's working *in vacuo*. Rather, accuracy in air remains stubbornly at a level not much better than half a part in a million, even after correcting or compensating for the variations of barometric pressure.

If this observation is correct, it would appear that the phasing of the drive may not after all be the factor that limits the accuracy of a pendulum swinging in air. There are other possibilities, such as one mooted by Bateman (1977, p. 57); the air itself may be responsible, not indirectly by reducing the Q and increasing the escapement error, but directly by the uneven way in which it itself obstructs the pendulum's motion. The disturbed air could act as a noise source closer to home than the rumble of distant traffic. The notion that air might absorb energy from the pendulum, yet feed noise back into it, may at first seem paradoxical, but there is nothing new in such an idea. In electrical tuned circuits, resistors act in just this way, simultaneously absorbing energy from the circuit and introducing thermal noise in the form of random voltage fluctuations. Closer to home, perhaps, we have only to think of a car with squealing brakes.

The possibility that a pendulum's timekeeping stability is limited by the turbulence of the air around the bob highlights the semantic distinction between noise and chaos. Noise refers to some external source of disordered energy entering our tidy world and upsetting things. Chaos – another word hijacked by science to be given a technical meaning – is different. It has recently been discovered that quite simple non-linear dynamical systems, even with as few as three degrees of freedom, can behave in a noisy manner spontaneously, giving the illusion of random behaviour whilst actually obeying the deterministic Newtonian laws of mechanics. This has caused a great intellectual stir, for we may no longer be forced to blame all randomness on the gods. The weather, although far removed from a simple system, is often quoted as an example of chaotic behaviour, but in spite of its unpredictability it is only responding to the energy of the sun in accordance with the laws of physics. Could it be that the air in a clock case responds to the swishing of the pendulum in an equally unpredictable manner? The alternative hypothesis would be that it settles down to a periodic and harmless pattern of streamlined flow almost impossible to imagine. The matter could surely be settled by experiment. By introducing smoke, Bateman has photographed the eddies around a pendulum bob and did at one time (Bateman 1977) attempt the crucial experiment of comparing the stability of a free pendulum with that of a driven one, but on examination I was far from convinced that his results were statistically significant.

In spite of this theory, my own hunch is that air around the pendulum is not what places a limit on the long-term stability of a clock working at atmospheric pressure. I find it hard to believe that chaotic air in a clock case could give rise to noise-like behaviour down to the very low frequencies present in flicker noise. If this is not wishful thinking on the part of a pendulum enthusiast, it may even yet prove possible to match the performance of a Shortt clock without a high vacuum. Rawlings was of the same opinion, though for reasons I cannot bring myself to accept (Woodward 1994). He was not persuaded that a high Q was a good thing.

The ultimate limit on pendulum accuracy must of course be set by fluctuations in the earth's gravitational force. These are of two kinds, the systematic – which might in theory be allowed for – and the unpredictable. That a clock pendulum can be made sensitive enough to detect systematic fluctuations of gravity was shown with the utmost clarity by the experiment carried out by Pierre Boucheron in 1984–1985. This experiment has already been well reported by Boucheron himself (1987), and also in Rawlings (1993, pp. 156–9), but as I was fortunate enough to participate in the analysis of the results, I cannot forbear to mention it again here.

It was an experiment waiting to happen, and the right circumstances arose when Boucheron, always the keenest of experimental horologists, saw his opportunity in a clock vault of the US Naval Observatory, Washington DC. There was Shortt No. 41, lying unused with its vacuum intact, and in a building with atomic time laid on. Out of sheer curiosity, Boucheron wished to find out how accurate the Shortt clocks had been, for when they were used as primary standards, this was never known with certainty. They could only be compared with other similar clocks and with the Earth, which is itself an unstable timekeeper. The Naval Observatory gave Pierre, then retired from his work at General Electric, a free run of the clock vault and direct access to Universal Time for the duration of his experiment, which was to run for some 352 days.

No modifications were made to the free pendulum of SH41, though the slave clock was replaced by an electronic slave more accurate than the free pendulum itself. This was an 'improvement' that may have been of little consequence, as it did not alter the fact that the time by SH41 was still that obtained from the free pendulum. Its time was compared with atomic time at hourly intervals, and the resulting data, consisting of some 8448 carefully averaged readings, were recorded to five decimal places of a second. The rate chart at figure 14.4 was obtained by differencing the error at daily intervals, but Boucheron also discovered a wealth of information hidden in the intervening hours. There the pendulum could be seen to have responded to the diurnal and semi-diurnal fluctuations of gravity caused by Earth tides. Reading his original account in *Antiquarian Horology* (Boucheron 1986), I determined to obtain from him the raw data, so that I could analyse it myself. Boucheron had been able to pick out tidal patterns without using any sophisticated methods, but the observations cried out for Fourier analysis, which I had used extensively in my professional work. My request for a copy of the data was gratifyingly answered by return of post, and I set about keying them in to my early home computer, a task that took an age and wore out the machine's flimsy keyboard. To save time, I replaced the entire computer.

A program for Fourier analysis had to be specially written, and being in a hurry I used the most obvious and therefore the slowest algorithm. So slow was the computer that the calculation had to be restricted to a couple of bands close to one and two cycles a day, where the tides are known to be. This might have pre-empted any sense of achievement when the spectral lines eventually showed up, but I had not expected to be confronted with as many as seven clearly resolved peaks. These were at frequencies of 0.929, 0.997, 1.003, 1.895, 1.932, 2.000 and 2.005 cycles per day, all apparently well known to geophysicists, and named O_1, P_1, K_1, N_2, M_2, S_2 and K_2. For reasons which I but dimly understand, the Sun and the Moon force our flexible ball of Earth to vibrate at numerous sharply defined anharmonic frequencies. To have a Shortt clock reveal this information for the first time on the monitor screen in my own study was a very special experience.

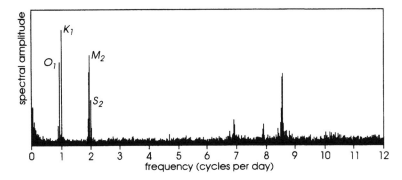

FIGURE 16.1
Frequency spectrum of Shortt clock rate, revealing the so-called larger lunar wave O_1, the lunisolar declinational wave K_1, the principal lunar wave M_2, and the principal solar wave S_2.

Later, using what is called the FFT (Fast Fourier Transform) in a much faster computer, I was able to produce the entire spectrum in about half a minute, Figure 16.1. The spectrum stops at 12 cycles per day for the simple mathematical reason that vibrations with periods shorter than two hours cannot be unambiguously detected in observations made at hourly intervals. On the scale of this figure, the fine structure of the spectrum cannot be seen, but only when the spectrum is viewed on this wide scale do we come across an unexpected hump at about 8.5 cycles per day, well separated from the tidal frequencies. Boucheron has informed me that this is the primary resonance of the planet itself, excited by chaotic processes such as volcanoes, earthquakes and storms.

The base line beneath the spikes and humps of the spectrum consists entirely of noise, which may or may not be gravitational in origin. The nature of the ultimate limit was the question which Professor Henry Wallman addressed in one of his last contributions to horology (Wallman 1992). In this, he appears to have been following up a suggestion by Bateman (1989) that rainfall might hold the key to the situation.

Wallman opened his paper with the observation that ever since Harrison's day the accuracy of pendulum clocks seems to have been limited to about one part in ten million – give or take a factor of four either way. At one extreme this range would embrace the late Harrison regulators, which were said to be good to a second a month, and at the other the Shortt clocks. In Wallman's opinion, the superior performance of observatory clocks such as the Shortt could be largely attributed to the constant temperature and pressure at which they were operated. By contrast with one part in ten million, quartz oscillators can achieve about one part in ten million million. The fact that after 250 years the pendulum had made such little progress in closing such an immense (potential) gap suggested the presence of some quite fundamental limitation on its performance.

Because frequency of vibration is proportional to the square root of the force of gravity, the fractional changes of gravity we are looking for amount to about two parts in ten million. Are such variations likely? To give us a feel for gravity fluctuations, Wallman shows us how much the force of gravity can change with position. At a latitude of 45°, moving 100 m closer to the equator decreases *g* by one part in ten million. More surprising still, raising a clock 12.5 inches higher on the wall weakens gravity by that same amount.

However, clocks are not affected by spatial fluctuations but by temporal ones, and here scientific knowledge seems sketchier. Wallman speculates that the most likely mechanism for seasonal fluctuations of gravity would be the movement of underground water, and he carries out several specimen calculations based on the work of Sir Harold Jeffreys. For example, if a horizontal porous rock stratum of thickness L should change its density by D, the change in gravity at the surface is proportional only to L and D and to Newton's universal gravitational constant. The formula is independent of the depth of the stratum, but its lateral extent is assumed much greater than its thickness, so the deeper it is, the more extensive it has to be. To cause a change of two parts in ten million in the force of gravity, Wallman assumes a density variation equal in magnitude to one quarter the density of water and calculates that the thickness of the porous slab need then be no more than 20 m. Porous rock or underground caverns causing seasonal variations in water content of 25% over a depth range of 20 m are all we need to bring about the longer term variations observed in pendulum timekeeping. It is a persuasive argument, and if correct it would have to be the last word.

CHAPTER SEVENTEEN

Clockwork with a difference

The 1860s saw the appearance of two most unusual regulator clocks, one in New England and one in Scotland. Both clocks were new to me until recently, both were eye-openers and I can think of no more pleasant way to finish this book than by giving a brief description of each. Their inventors were William Bond of Messrs Bond and Son, Boston, Massachusetts, and Lord Kelvin (then Sir William Thomson) of the University of Glasgow. By the merest chance, only three clocks of each type were made.

Two of the three by William Bond are still on home ground in the USA, but one is in Liverpool, England, having been ordered especially for the Observatory there by its director John Hartnup in the 1860s. It is now in the possession of the Liverpool Museum, where it can be seen in the new Space and Time Gallery. The Bond regulator has a single-beat gravity escapement, but the surprise feature is that detachment of the pendulum from the weight-driven train is *complete*. This was the problem that had so exercised the mind of Captain Henry Kater in the years up to his death in 1835. I have shown in chapter 3 how James Arnfield in 1986 succeeded in carrying Kater's idea to a successful conclusion, but William Bond's solution was entirely different.

To recapitulate, the problem with a gravity escapement has always been that of resetting the gravity arm or arms without requiring the pendulum to do any of the work. Some care is needed in the statement of this problem. Obviously, it is out of the question for the pendulum to do the actual resetting, because this would take as much energy as the gravity arms had supplied in the first place. The actual work must be done by the clock train, which lies dormant until unlocked for the purpose. The problem concerns the unlocking, which must of course consume a small amount of effort. Where is this effort to come from?

In Grimthorpe's otherwise admirable escapement, the pendulum itself does the work by sliding the locking pallet on the gravity arm away from the escapement leg that has been

Regulator clock No. 395 by William Bond in the Space and Time Gallery of Liverpool Museum, England. The clock is powered by two spherical weights whose fall is governed by the conical pendulum under the dome above the dial. A similar clock (No. 394) can be seen in the Harvard University Collection of Historical Scientific Instruments.

pressing upon it. This is about to happen in the left-hand drawing of Thwaites and Reed's variant of Grimthorpe's escapement in Figure 3.8. The objection is not so much that the pendulum is being subjected to frictional drag but that the force of the drag depends directly upon the considerable and possibly variable pressure being exerted on the locking pallet by the clock train. Grimthorpe's contribution was to minimize the drag by using a wheel with long legs in place of a conventionally toothed escape wheel.

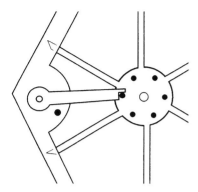

FIGURE 17.1
Detail from Figure 3.9, showing the safety lip in Arnfield's gravity escapement.

Ideally, of course, there would be no load at all on the pendulum. For a *central* gravity impulse, this appears to be a logical impossibility. Even Shortt's free pendulum suffers a small preliminary drag, but the very mention of impossibility will, I hope, tempt some enthusiastic horologist to prove me wrong. Certainly it should be provable one way or the other. In 1984, John Wright exhibited an entertaining clock which solved the problem completely by arranging for a pallet on his free pendulum to intercept a stream of water (Wright 1977), but I know of no solution in terms of pure mechanics. For the conventionally *recoiling* gravity escapement, a little extra loading can hardly matter, as the pendulum is already having to lift the gravity arm, but it remains important that this extra loading should be independent of the force from the train.

Arnfield's solution has already been discussed in chapter 3. He almost eliminates the loading, though it is to be noticed that when the pendulum lifts the gravity arm, it suffers a tiny dragging force from a safety lip, Figure 17.1. This force does not originate from the clock train, and by the use of light materials can be made almost as small as may be desired. Pure mathematicians are well accustomed to the notion of a quantity that is greater than zero but as small as may be desired, and here the same concept crops up in horology. We need a term for it.

William Bond invented what is, in effect, a *mechanical relay*, that is to say a mechanism making no direct contact with the power train, yet able to bring it into action with the tiniest force when required. Above the movement and dial there is a conical pendulum, whose bob rotates in a circle in the horizontal plane. The beauty of a conical pendulum, Figure 17.2, is that its motion is continuous and can be maintained without any kind of tick-tock escapement. It is not a good timekeeper in itself, because the greater the torque applied to the drive shaft the faster it will go. Even a brake in the form of an embracing ring to limit the rise of the bob will not ensure sufficient accuracy for precision timekeeping. William Bond uses the conical pendulum to keep his clock train in continuous motion, but the speed of rotation is not highly critical.

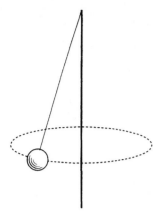

FIGURE 17.2
Conical pendulum.

Below the movement is the real pendulum, which takes impulse from a single gravity arm. From the viewpoint adopted in Figure 17.3, the pendulum has received impulse on its leftward swing, and has left the gravity arm on a banking in the normal manner. At the moment shown in Figure 17.3a, a wheel I have called the *remontoire wheel* is on its way to reset the gravity arm in readiness for the next impulse. The remontoire wheel is driven by the continuously running clock train, and the gravity arm is lifted by a hemi-cylindrical pin pushing against a disc pivoted on the arm. The remontoire wheel and its driving wheel are shown here with smooth rims, and those of the Liverpool clock are indeed

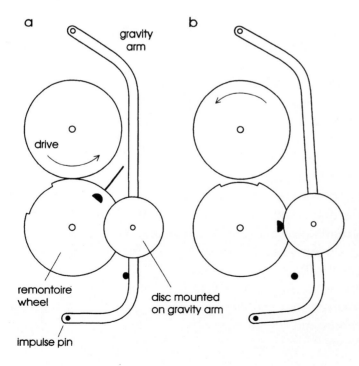

FIGURE 17.3
William Bond's gravity escapement in schematic form.
(a) The pendulum rod swinging left (not shown) has received impulse, leaving the gravity arm at its banking.
(b) The gravity arm has been reset, awaiting return of the pendulum.

smooth. The weight of the driving wheel (one of whose pivot holes is slightly over-sized) ensures a sufficient grip, but all three of these clocks originally had toothed wheels. It appears that Hartnup had the teeth removed shortly after the clock was delivered to him at Liverpool.

The reset is complete when the gravity arm has been lifted as far as the disc allows, as shown in Figure 17.3b. Here the remontoire wheel is brought to rest as its finger comes up against a stop on the gravity arm, not shown in the drawings. The wheel is now disengaged from the train because of a short cut-away portion on its rim. However, it is still subject to a torque because it is deliberately made out of poise, as though from a weight attached to its finger.

When the pendulum returns from its left excursion and picks up the gravity arm, the stop that had detained the finger of the remontoire wheel is pulled away, leaving the wheel free to start turning once again. The resistance offered to the pendulum by this unlocking action is not only small but is completely independent of the drive train.

The final stage is the re-engagement of the remontoire wheel with the driving wheel. Even with teeth the shock is not great, as the teeth will already be moving in the same direction. It is a little like running to board a moving train.

All my information about the Bond regulator has been gleaned from a scholarly account given by John Griffiths (1987) and illustrated with true-to-life drawings by David Penney in *Antiquarian Horology*. My own drawings are a gross distortion of the actual mechanism, and are not intended to be lifelike. I have used mathematical licence to help make the principle of the escapement clear.

The principle of the relay is most familiar to engineers in an electrical setting, such as the electric starter switch of an automobile. Is there anything to be learned about the borderline between mechanical and electrical techniques from William Bond's regulator? The facility offered by an electromagnet, as for example in the resetting of the Synchronome clock's gravity arm, is the ability to switch on a large force simply by placing one piece of metal in contact with another. The arrival of some given moving part at the appropriate place triggers off a powerful action. In an attempt to imitate the Synchronome switch, the gravity arm in my W5 clock is reset when it contacts the escape wheel, but it has to *jolt* it to wake it up. In the Bond escapement, we have a genuine relay action, in which the remontoire wheel has only to sidle up to the drive wheel to be caught up in a determined movement. One might have imagined this to be impossible in pure mechanics, but it is clearly not so. Interestingly, the same technique of a cut-away wheel was used in the remontoire of the Selticon® clock made by Portescap of Switzerland in comparatively recent times. Its resonator was a balance and hairspring, impulse being given through a remontoire spring which was reset by a continuously running transistor motor.

It would be futile to suppose that mechanics is in some sort of competition with electricity, whose real magic resides in electric and magnetic fields of force. Mechanics has its own field of force, namely that of gravity, which is less pliable. On the other hand, the human ingenuity used in exploiting it seems to know no bounds.

In a comprehensive paper published in *Antiquarian Horology* and in a more recent paper, Charles Aked (1973, 1994) describes a clock designed by Lord Kelvin (as I shall call him hereafter) as the 'first free pendulum clock'. Opinions may differ as to whether it truly merits the title 'free', but the name does not matter.

Kelvin's clock shares one outstanding feature with that of William Bond, which is the use of a continuously running motor, but there the resemblance ends. As may be seen from the photographs, Kelvin's clock has (like W5) two pendulums in one case. Three copies of the clock are said to have been made, one of which is now in working order in what was Kelvin's residence at the University of Glasgow. A second version, after a checkered history, is now in the possession of the London Science Museum, and the third – if it ever existed – seems to be lost. Kelvin never attempted to bring his clock up to observatory pitch, and we do not yet know how well it would have performed in practice.

To understand the principle of this extraordinary clock we can forget until a later stage that it has two pendulums. At Figure 17.4a I have drawn an ordinary dead-beat escapement with pallets at diametrically opposite positions in relation to an escape wheel equipped with but a solitary tooth. This is equivalent to the geometry of Kelvin's actual escapement sketched at Figure 17.4b, where everything is turned on its side under the bob. The dead-beat pallets are mounted on a frame fixed to the base of the bob, and the escape wheel is reduced to a collar on a vertical arbor which I shall call the *drive shaft*. The single tooth of the escape wheel takes the form of a wire whirling round horizontally. After giving impulse, the wire does indeed whirl, but not in an uncontrolled manner as in the drop of an ordinary escapement. The speed of rotation of the wire is under the control of a continuously running and carefully adjusted centrifugal governor, so that after escaping from one pallet as the pendulum is in the middle of its leftward swing, it will not reach the other pallet until the pendulum is ready for it in the middle of its swing towards the right.

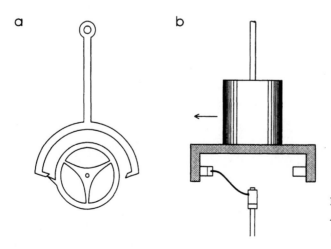

FIGURE 17.4
*Kelvin's escapement,
(a) conceptually and (b) actually.*

The second of Lord Kelvin's free pendulum clocks, now in the keeping of the London Science Museum.

For this idea to work properly, the scape collar (as I shall call it) must be able to pause momentarily whilst the wire is in contact with the dead face of the pallet, and during the impulse it must of course proceed at a speed dictated by the angle of the pallet and the velocity of the pendulum. Once the wire has escaped, however, the collar can resume the speed dictated by the governor, which must run a little faster than the frequency of the pendulum in order to make up for lost time.

Here we have a problem. To most horologists it will seem strange that it should have been solved in the way it was, which was to fit the scape collar loosely on the drive shaft, and rely on *friction* for its propulsion. During impulse, slippage occurs between the collar and the shaft, so the actual force of the impulse is entirely dependent on friction, a matter to which I shall return later. Meanwhile, let us take stock of what Kelvin has now achieved. The pendulum swings freely to its extremes of arc, receiving impulses as it passes through centre automatically. It has no recoiling of a gear train to endure, as in the anchor escapement, no friction rest at the extremes of swing, as in the Graham, and indeed no unlocking of the train to perform, as in the detached lever escapement of a watch. These are the characteristics which qualify the pendulum as 'free'.

The radical departure from conventional horology is of course the independence of the drive. In an ordinary clock escapement, the speed of the train is locked to the frequency of the pendulum, which is why the train can be used to register the time. In Kelvin's clock (as thus far described) the drive shaft is no more accurate as a timekeeper than the centrifugal governor of which it forms a part. The time kept by the pendulum has somehow to be taken from the motion of the scape collar.

It appears that Kelvin had some difficulty in making a governor accurate enough for his purpose. In its later form it consisted of brake shoes hanging like a pair of scales from the ends of a beam fixed to the drive shaft. As the shaft rotated, the shoes would tend to fly out by centrifugal force but were constrained by a circular hoop. The faster the rotation, the greater the force on the hoop and the greater the braking effect, causing the system to settle down to a stable angular velocity. In more recent times this same centrifugal principle was used to control the speed of wind-up gramophone turntables which could keep music in tune to a fraction of a semitone, but although Kelvin had succeeded in making a clock along these lines before 1865, he was not satisfied.

The next innovation was to use the pendulum itself for fine control of the governor. Like the house that Jack built, the pendulum controlled the governor that governed the speed of the shaft that drove the collar with the wire that impulsed the pendulum! This is pure cybernetics. I shall condense the explanation into one optional paragraph for those readers who are not to be intimidated, and I make no apology for the absence of a drawing. I believe this is one of those rare cases in which principles can be more easily understood from a verbal description alone.

The pendulum used for this mechanism was a half-second one (period 1.0 hertz). Two shafts were needed, the drive shaft as before, now made in the form of a pipe enclosing another shaft, the *time shaft*. As already explained, the drive shaft has to go a little faster (it actually went 4% faster) than the *average* speed of the scape collar, which was precisely one rotation per second, being controlled by the pendulum. By means yet to be described, the time shaft is made to turn smoothly at one revolution per second *in synchronism with the pendulum*. Assuming this can be done, the drive shaft can be geared from it so as to go faster by just that extra 4% needed to take the wire from one pallet to the next. In this new arrangement, the scape collar is still propelled by the drive shaft, friction being applied through leaf springs that allow the collar some vertical freedom of movement. The collar is located vertically by a pin engaging with a helical groove on the protruding time shaft. If the time shaft really does go at one revolution per second, all is well, because the scape collar is likewise turning at 1 rps on average. Thus it will not screw itself up or down the time shaft, except by a tiny amount each way twice per revolution. The servo-mechanism is completed by using the vertical position of the collar as the speed control for the governor. The collar then settles at the height required for synchronism automatically. Its control over the governor is effected by long slender coil springs attached to the governor's brake shoes, so altering their pressure on the constraining hoop. The springs can be seen in Figure 17.5 at 45° to the horizontal plane of the hoop. The governor has considerable inertia and will not be able to respond to the tiny speed fluctuations caused by the impulsing, but it will respond to the *average* speed of the time shaft, which is all that is required. The ingenious principle of this governor had earlier been communicated to Kelvin by Professor Fleeming Jenkin of Edinburgh University.

Finally, Kelvin used one of the now very accurately governed shafts, after further gearing, as the drive for another pendulum, a seconds pendulum (0.5 hertz) which can be seen with its larger bob to the left of the shorter pendulum. I shall call this the free pendulum.

FIGURE 17.5
Close-up of Lord Kelvin's clock. The shorter pendulum (right) swings from side to side; the free pendulum (left) swings towards and away from the observer. The small dials record seconds and minutes from the free pendulum.

It is impulsed on exactly the same principle by a slipping collar, but with no governor to control. The drive is geared to run a mere 0.25% faster than the pendulum, the idea being to shorten the impulses to something like 1/400 second. The amplitude of swing was reportedly ±0.5 cm at the bob, a semi-arc of less than one third of a degree, which is very small compared with that of ordinary pendulum clocks. The two pendulums swung in perpendicular planes, presumably to guard against any tendency of the swings of the short pendulum to influence the swings of the free pendulum via reactions transmitted through the supporting structure. Any such effect would vitiate the underlying object of the double pendulum concept, which was to make one oscillator do the donkey work for an independent oscillator of superior accuracy. The master and slave principle of Shortt's clock is here anticipated by half a century.

The large dial at the top of Kelvin's clock shows hours and minutes, and is driven from the time shaft of the shorter and less accurate pendulum. (The vertical rod communicating this motion can be seen clearly between the two pendulums in the close-up photograph.) The time given by the free pendulum is shown on a tiny seconds dial below the bob of the free pendulum, its hand being driven from the free pendulum's scape collar which is cut as a worm gear for the purpose. On the Glasgow clock, there is no further gearing from the seconds hand, which is thus the only register of really accurate time. As the two pendulums kept time independently, the main dial is of no use for resolving the ambiguity of sixty seconds, but this would not matter to astronomers, who are expected to know their minutes already! The clock illustrated in the photographs differs from the Glasgow clock in having an additional small dial for registering the minutes from the free pendulum, and there Kelvin drew the line.

One cannot but reflect on the likely performance of this extraordinary clock. When the impulsing wire escapes from a pallet, the collar will continue to slide on its shaft for a short while, for every mechanical object has inertia which stops it from acquiring a new velocity instantly. The extra slide makes the timing of the next contact with a pallet depend to some extent on friction, and seems certain to affect the stability of the pendulum's amplitude.

In spite of everything, my feeling is that the variability of frictional forces must be the most serious objection to Kelvin's escapement, but the small amplitude of swing, 0.3°, is its saving grace. Even if the amplitude were to vary, the effect on the timekeeping would be small. For a given *fractional* variation of amplitude, the variation of circular error is proportional to the square of the amplitude. With an amplitude five times smaller than is usual for a Graham regulator, therefore, circular error variations would be twenty-five times smaller. This is another justification for Kelvin's very short impulses. As a means of achieving very small amplitudes, short impulses may have been an easier solution than reduction of their force, which would involve reducing the frictional torque of the drive. The latter might present difficulties, because that same frictional torque has to be large enough to drive the counting mechanism. This undoubtedly explains why, on the Glasgow clock, the freer long pendulum has only a seconds hand to operate.

Kelvin (Thomson 1869) published an account of his clock in several scientific journals, starting with the *Proceedings of the Royal Society*, but the clock was never timed in earnest. He was well aware that accuracy could be limited by other factors, such as dimensional instability of materials. Prophetically he suggested that the pendulum should be totally enclosed in order to isolate it from barometric variations, and should be kept at a constant temperature throughout the year.

The clock was driven by a weight of 120 lb falling through twelve feet in a week, for which it was found necessary to cut a hole in the floor, through to the cellar below! Today we would use electrical rewinding, as Peter Brain has done in an interesting reproduction. He it was who first drew my attention to Charles Aked's research, and who demonstrated his working version of the clock even to the point of dismantling some of the critical components.

It is open to speculation whether Lord Kelvin knew of William Bond's activities when he was doing his early experiments in the years up to 1865. John Griffiths tells us that the instruction manual for the Bond regulator at Liverpool was despatched on November 5th, 1867. That two clocks using continuous motion should have been made at virtually the same time must be intriguing for scientific historians. As I am not of their number, I must leave this question, along with numerous others which will have been raised in these pages, unanswered.

Appendix: The centre of oscillation of a pendulum

Clockmakers need some way of calculating the period of vibration of a pendulum, T, in terms of its length, L. The standard formula is

$$T = 2\pi \left(\frac{L}{g}\right)^{\frac{1}{2}}, \tag{1}$$

but what exactly is the length L? It is the distance between the point of suspension and the *centre of oscillation*. For an idealized pendulum in which all the mass is concentrated at one point, the centre of oscillation is at the centre of gravity. The mass of a real pendulum, however, is distributed, and the centre of oscillation no longer coincides with the centre of gravity. The purpose of this appendix is to show that L is then equal to k^2/h, where k is the *radius of gyration* of the pendulum about its point of suspension, and h is the length (Figure A1) measured from the point of suspension to the centre of gravity G.

The pendulum's instantaneous angle with the vertical can be taken to be

$$\theta = \alpha \cos\omega t, \tag{2}$$

where α is the amplitude and ω is the angular frequency in radians/s. The *potential energy* of the pendulum at angle θ is

$$P.E. = mgh(1 - \cos\theta) \approx \frac{mgh\theta^2}{2}.$$

The *kinetic energy* at angle θ is

$$K.E. = \frac{mk^2\dot\theta^2}{2}.$$

APPENDIX

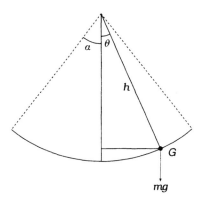

FIGURE A1
Diagram of pendulum at an arbitrary angle θ.

Using (2), the total energy is therefore

$$P.E. + K.E. = \frac{m\alpha^2(gh\cos^2\omega t + k^2\omega^2\sin^2\omega t)}{2}$$

$$= \frac{m\alpha^2\{gh - (gh - k^2\omega^2)\sin^2\omega t\}}{2}. \quad (3)$$

As the total energy must remain constant at all times, the coefficient of $\sin^2\omega t$ must be identically zero. This gives

$$\omega^2 = \frac{gh}{k^2}. \quad (4)$$

The period T is related to ω very simply, because the rate of change of phase, ω, times the period T, must be equal to 2π radians. Therefore

$$T = 2\pi/\omega. \quad (5)$$

Substituting for ω from (4),

$$T = 2\pi\left(\frac{k^2}{gh}\right)^{\frac{1}{2}}. \quad (6)$$

This, then, is the formula for the period of a pendulum in terms of k^2 and h. By comparison with equation (1), we see that $L = k^2/h$, which is what we set out to show.

Glossary

amplitude
Of a pendulum or balance, the angle of swing from the centre to either extreme. The extent of any sinusoidal variation (e.g. of force or velocity) measured from zero to maximum.

anisochronous
Lacking isochronism, i.e. having a period dependent on amplitude.

anti-phase
Two vibrations are in anti-phase if they have the same frequency but are out of step by exactly half a period, e.g. two seconds pendulums passing through centre in opposite directions at the same moment.

arbor
Horological term for a rotating shaft or axle.

armature
A part which moves in response to a magnetic force, typically from an electromagnet.

backstop
A *pawl* preventing reverse motion of a ratchet wheel.

balance
Inertial element in a sprung resonator, e.g. the vibrating wheel in a mechanical watch, to the arbor of which one end of the hairspring is connected (see Figure 10.1). See also *balance wheel*.

balance wheel
Ambiguous term used casually to mean the *balance* of a watch. In a verge escapement, however, the balance wheel is the *escape wheel* that drives the balance.

GLOSSARY

banking
A pin, peg, or other means of limiting the extent of movement of any component.

barrel
In clockwork, the cylindrical reel which directly or indirectly drives the train, either by having the line for a hanging weight wound around it or by having a mainspring wound inside it.

bob
The weight on the bottom of the pendulum rod, so called because it once bobbed in and out of the case.

centre of oscillation
The point at which all the mass of a pendulum could in theory be concentrated without altering the period of vibration (see Appendix).

circular error
Of a pendulum, the small change of *rate* resulting from a change of amplitude, caused by the circular path of the bob.

click
Horological term for a *pawl*.

coefficient
The amount by which something changes in response to unit change of some other stated quantity such as temperature or pressure.

compliance
Amount of elastic 'give' per unit force applied, e.g. a soft chair has a large compliance.

composer
Part of a grasshopper escapement causing the elbow joint to relax to a fixed angle (see Figure 5.2).

count wheel
Ratchet wheel used in counting vibrations. (The term is also used for the locking plate in a striking clock governing the number of strikes at the hour.)

crown wheel
An *escape wheel* whose teeth are perpendicular to the plane of the wheel. See Figure 1.3.

crutch
A fork engaging with a pendulum rod to deliver impulse. See Figure 3.4.

damping force
Force which, by resisting motion, causes natural vibrations to decay in amplitude. A damping force proportional to velocity produces exponential decay.

detent
A part designed to hold back the movement of a wheel, typically the escape wheel.

GLOSSARY

drop
 Brief runaway motion of an escape wheel after a tooth has escaped from a pallet and before it re-locks.

escape wheel
 Toothed wheel engaging with pallet(s) to drive a balance or pendulum, either directly or through some intermediate mechanism.

exponential
 Pattern of growth or decay whose rate is proportional to the magnitude of the quantity that is growing or decaying.

fall
 Depth available for the fall of a driving weight.

flicker noise
 Sometimes called $1/f$ noise; a continuous random fluctuation whose power is spread equally over every octave of frequency.

fly
 Fan or flywheel to control speed of rotation.

foliot
 The inertial element used in ancient clocks, see Figure 1.3.

Fourier analysis
 The analysis of any periodic time variation into sinusoidal components at multiples of the fundamental frequency. (Named after J. B. J. Fourier.)

frequency
 Number of vibrations per second; if the period is T seconds, the frequency is $1/T$ hertz; similarly if the frequency is f hertz, the period is $1/f$ seconds.

fusee
 A tapered, spirally grooved, pulley intervening between the mainspring and the gear train of a spring-driven timepiece. The fusee is driven by a chain (or gut line) from the spring barrel. Its varying diameter compensates for variation of torque from the mainspring.

going train
 Sequence of wheels and pinions conveying power from the barrel to the escapement of a clock; the term contrasts with the striking train, which delivers power to the striking mechanism.

gravity arm
 Pivoted arm, which, after having been lifted mechanically or electrically, delivers impulse to a pendulum by its weight alone.

hertz
SI unit of *frequency* equal to one vibration per second.

homing force
Synonymous in this book with *restoring force*.

hysteresis
A lag which causes a system to respond differently according as the applied stress is increasing or decreasing.

impulse
Regular push given to a balance or pendulum to maintain it in oscillation. As a verb, 'to impulse' means 'to give impulse'.

isochronism
Period of vibration not varying with amplitude.

jockey pulley
A pulley to take up the slack in a line or belt.

lantern pinion
A pinion constructed from a pair of discs connected by rods, as in a Davy lamp. See Figure 1.2.

line
Gut, rope, cable or chain on which the driving weight of a clock is suspended.

millibar
Measure of barometric pressure; one bar is almost exactly 750 mm of mercury.

moment of inertia
Rotational inertia; mathematically mk^2, where m is the mass and k is the radius of gyration.

motion work
Wheel work which transforms speed of rotation without conveying power, as used to gear the hour hand of a clock from the minute hand.

noise
Unwanted random fluctuations disturbing some wanted time-varying process.

pallet
Typically a pad designed to transmit impulses from an escape wheel to a balance or pendulum.

pawl
Pivoted arm or lever catching on the teeth of a ratchet to prevent reverse motion or, by its own movement, to move a ratchet onwards. In horology, the word more commonly used is *click*.

period
  Time taken for one complete cycle of vibration, e.g. twice the time taken for a pendulum to swing from one extreme to the other.

phase
  Stage reached in a vibration, measured in terms of an angle advancing by 360° (2π radians) per period.

pinion
  Small solid gear wheel meshing with a larger gear wheel.

ppm
  Parts per million.

Q
  Dimensionless number measuring a resonator's freedom to vibrate.

quadrature
  Two vibrations are in quadrature if they have the same frequency but their phase angles differ by 90° ($\pi/2$ radians).

random walk
  Aimless travel in one or more dimensions resulting from a succession of random steps.

ratchet
  A rack, or more usually a wheel, with saw-shaped teeth to be engaged by a *pawl* or *click*.

rate
  The amount by which a clock gains or loses in unit time. In horology, rate is usually averaged over a day and measured in seconds per day. Scientifically it is more conveniently expressed in parts per million, 1.0 ppm corresponding to 0.0864 s/day. Modern usage reckons a gaining rate as positive and a losing rate as negative, but in astronomy the opposite convention was used.

regulator clock
  Accurate pendulum clock such as would be used for astronomical observations. Later, regulators were used as a shop's time standard for regulating other clocks and watches.

remontoire
  Literally a rewinding mechanism, but usually taken to mean a subsidiary drive, periodically rewound by the main power train of a watch or clock, with the object of supplying a more constant drive to the succeeding mechanism.

resonator
  Any system capable of vibrating at a discernible frequency.

restoring force
  The force which attempts to return a resonant device to its rest position.

GLOSSARY

scape (or 'scape)
   Traditional abbreviation for *escape wheel*.

semi-arc
   Angle of swing of a pendulum from centre to extreme, i.e. its amplitude.

SHM
   Simple harmonic motion, the sinusoidal form of vibration resulting from a restoring force proportional to displacement from a central point.

sinusoid
   The shape of the graph of sin x when plotted against x.

solenoid
   A coil of wire which produces a magnetic field when current is passed through it; an electromagnet.

sprocket
   A wheel with teeth or spikes to engage with the links of a chain.

staff
   Another word for an axle; traditionally, the arbor of a balance is called the *balance staff* and the arbor for the pallets of a watch escapement the *pallet staff*.

supplementary arc
   Excess amplitude of swing beyond the minimum required for the escapement to operate (the *escaping arc*).

swing
   Usually taken to mean half a vibration.

train
   Cascade of gear wheels meshing with pinions, typically in clockwork to amplify motion by a large factor at the cost of reduced torque.

verge
   Pallet arbor of the verge escapement (see Figure 1.3).

vibration
   One complete cycle of a resonator's motion; the time taken for one vibration is one *period*.

wheel
   In clockwork, a 'wheel' is understood to mean a toothed gear wheel. (For this reason the term *balance wheel* should be avoided when describing the *balance* of a mechanical watch or clock.)

white noise
   A continuous random fluctuation whose power is spread uniformly over the entire spectrum of frequencies, by analogy with white light.

References

Airy, G. B. (1827). On the disturbances of pendulums and balances, and on the theory of escapements. *Transactions of the Cambridge Philosophical Society*, **3**, 105–128.

Airy, G. B. (1829). On a correction requisite to be applied to the length of a pendulum consisting of a ball suspended by a fine wire. *Transactions of the Cambridge Philosophical Society*, **3**, 355–360.

Aked, C. K. (1973). The first free pendulum clock. *Antiquarian Horology*, **8**, 136–162.

Aked, C. K. (1994). The first free pendulum clock. *Bulletin of the Scientific Instrument Society*, **41**, 20–23.

Allan, D. W., Shoaf, J. H., and Halford, D. (1974a). Statistics of time and frequency data analysis. In *Time and frequency: theory and fundamentals*. NBS Monograph 140 (ed. Blair, B. E.), pp. 151–204. US Department of Commerce, Washington DC.

Allan, D. W., Gray, J. E., and Machlan, H. E. (1974b). The National Bureau of Standards atomic time scale: generation, stability, accuracy and accessibility. In *Time and frequency: theory and fundamentals*. NBS Monograph 140 (ed. Blair, B. E.), pp. 205–231. US Department of Commerce, Washington DC.

Arnfield, J. (1987). An inertially detached gravity escapement. *Horological Journal*, **130** (4), 10–12.

Baillie, G. H., Ilbert, C., and Clutton, C. (1982). *Britten's 'Old clocks and watches and their makers'* (9th edn. revised Clutton). Methuen in association with E. & F. N. Spon, London.

Bateman, D. A. (1977). Vibration theory and clocks, part 5: Effects of external vibrations. *Horological Journal*, **120** (5), 52–7.

Bateman, D. A. (1989). The accuracy of pendulum timekeepers, part 4. *Horological Journal*, **132**, 203–4.

Bateman, D. A. (1993). Letters. *Horological Journal*, **136**, 112.

Bateman, D. A. (1994). Accuracy of pendulums and many factors that influence it. *Bulletin of the National Association of Watch and Clock Collectors*, **36**, 300–312.

REFERENCES

Beeson, C. F. C. (1957). A history of the Wadham College clock. *Antiquarian Horology*, **2**, 47–50.

Betts, J. (1993). *Harrison*. National Maritime Museum Publications, London.

Boucheron, P. H. (1986). Effects of the gravitational attraction of the sun and moon on the period of a pendulum. *Antiquarian Horology*, **16**, 53–65.

Boucheron, P. H. (1987). Tides of the planet Earth affect pendulum clocks. *Bulletin of the National Association of Watch and Clock Collectors*, **29**, 429–33.

Britten, F. J. (1938). *The watch and clock makers' handbook, dictionary and guide* (revised by Player). E. & F. Spon, London and New York.

Brown, E. W. and Brouwer, D. (1931). Analysis of records made on the Loomis chronograph by three Shortt clocks and a crystal oscillator. *Monthly Notices of the Royal Astronomical Society*, **91**, 575–591.

Cain, D. and Boucheron, P. (1994). Circular error defeated at last. *Horological Science Newsletter*, National Association of Watch and Clock Collectors (Chapter 161), Issue 1994-1.

Chamberlain, P. M. (1941). *It's about time*. Holland Press, London.

Cuss, T. P. C. (1965). The Huber/Mudge timepiece with constant force escapement. *Pioneers of precision timekeeping*, Monograph No. 3 of the Antiquarian Horological Society, 93–115.

Gazeley, W. J. (1980). *Clock and watch escapements*. Newnes Technical Books, London.

Gould, R. T. (1960). *The marine chronometer – its history and development* (revision of 1923 edition). The Holland Press, London.

Green, E. I. (1955). The story of Q. *American Scientist*, **43**, 584–595.

Griffiths, R. J. (1987). William Bond astronomical regulator No. 395. *Antiquarian Horology*, **17**, 137–144.

Grimthorpe, Lord* (1903). *A rudimentary treatise on clocks, watches and bells for public purposes* (8th edn). Reprinted in 1975 by E. P. Publishing, Wakefield.

Grollier de Servière (1719). *Recueil d'ouvrages curieux de Mathématique et de mécanique etc.*, David Forey, Lyon.

Hastings, P. (1993). A look at the grasshopper escapement. *Horological Journal*, **136**, 48–53.

Hope-Jones, F. (1949). *Electrical Timekeeping* (2nd edn.). N. A. G. Press, London.

Howse, D. (1980). *Greenwich time and the discovery of longitude*. Oxford University Press.

Howse, D. and Hutchinson, B. (1971). The Tompion clocks at Greenwich and the deadbeat escapement. *Antiquarian Horology*, 7, 18–34 and 114–133.

Jespersen, J. and Fitz-Randolph, J. (1982). *From sundials to atomic clocks*. Dover Publications, New York.

---

* The English practice by which names change with promotion can be confusing, especially in this case. Lord Grimthorpe was originally Edmund Beckett Denison (the name Denison having been adopted by his father in 1816). In 1874, E. B. Denison succeeded to his father's title of Baronet, dropped the name Denison and became Sir Edmund Beckett, Bt. In 1886, however, he was created 1st Baron Grimthorpe, and it is as Grimthorpe that we now usually refer to him.

REFERENCES

Kater, E. (1840). Description of an escapement for an astronomical clock, invented by the late Captain Henry Kater, FRS etc. *Philosophical Transactions of the Royal Society*, **130**, 335–340 and Plate XI.

Laycock, W. (1976). *The lost science of John 'Longitude' Harrison*. Brant Wright Associates, Ashford.

Lepaute, J. -A. (1755). *Traité d'horlogerie*. Jacques Chardon, Paris.

Mandelbrot, B. (1967). Some noises with $1/f$ spectrum, a bridge between direct current and white noise. *Transactions on Information Theory*, Institute of Electrical and Electronic Engineers, **IT-3**, 289–298.

Newton, I. (1687). *Philosophiae Naturalis Principia Mathematica* (trans. Motte, 1803).

Rawlings, A. L. (1993). *The science of clocks and watches* (3rd edn, ed. Treffry, Timothy and Amyra). British Horological Institute, England.

Rees, A. (1820). *Clocks, watches and chronometers*. David and Charles Reprints (1970). David and Charles, Newton Abbot.

Riefler, D. (1981). *Riefler-präzisionspendeluhren 1890–1965*. Callwey Verlag, Munich.

Royer-Collard, F. B. (1969). *Skeleton clocks*. N. A. G. Press, London.

Sampson, R. A. (1928). Studies in clocks and timekeeping: No. 4. The present-day performance of clocks. *Proceedings of the Royal Society of Edinburgh*, **48** Part 2, 161–166.

Shelley, F. (1987). Aaron Dodd Crane, an American original. *Bulletin Supplement 16*, National Association of Watch and Clock Collectors, Columbia PA.

Shortt, W. H. (1929). Some experimental mechanisms, mechanical and otherwise, for the maintenance of vibration of a pendulum. *Horological Journal*, **71**, 224–226 and 245–247.

Strong, C. L. (1960). How two distinguished amateurs set about refining the accuracy of a pendulum clock. *Scientific American*, **203** (1), 165–176, and (2), 158–168.

Ta'Bois, N. C. (1984). The first clock kit? *Clocks*, 7 (4), 45–6.

Thomson, W. (1869). On a new astronomical clock, and a pendulum governor for uniform motion. *Proceedings of the Royal Society*, **17**, 468–70.

Wallman, H. (1992). Do variations in gravity mean that Harrison approached the limit of pendulum accuracy? *Horological Journal*, **135**, 24–25.

West, B. J. and Shlesinger, M. F. (1989). On the ubiquity of $1/f$ noise. *International Journal of Modern Physics B*, **3**, 795–819.

West, B. J. and Shlesinger, M. F. (1990). The noise in natural phenomena. *American Scientist*, **78** (40–45).

Woodward, P. M. (1974). A new master and slave system. *Horological Journal*, **117** (3), 3–8.

Woodward, P. (1989). A new look at escapement theory. *Horological Journal*, **131** (10), 9–11, (11), 25–27, (12), 12–13, **132** (1), 16–17 and 27, (2), 46–48 and p. 95 (correction).

Woodward, P. (1993). The performance of a 19th century regulator. *Horological Journal*, **135**, 306–12.

Woodward, P. (1994). Air pressure, impulse, and the Shortt clock. *Horological Journal*, **136**, 368–9.

Wright, J. F. (1977). Free pendulum clock, with liquid escapement. *Horological Journal*, **120** (5), 7–10.

Acknowledgements

The author and the publisher are grateful to the following for supplying photographs and for permitting their reproduction in this book.

Oxford University Museum of the History of Science: for the photograph on page 22.

Trustees of the British Museum: for the photograph on page 53.

Prescot Museum of Clock and Watchmaking, and National Museums & Galleries on Merseyside: for the photographs on pages 86 and 140.

Mr Geoffrey Goodship: for the loan of colour negatives from which the photographs on pages 145 and 147 were prepared.

Index

Page numbers in bold face refer to photographs, those in italics to graphs or diagrams. The abbreviation n. refers to a footnote or a caption.

Accutron watch 8
adopted error 109
air resistance 12, 135
Airy, Sir George Biddell
  (1801–1892) 63–5, 69, 83,
  104
Airy's laws 63–4
Aked, Charles K. 144
Allan, David W. 123, 130
anchor escapement 21–3, **22**
  used in reverse 43–4, *44*
aneroid capsule 102–4
anisochronism 9, 11, 36
Archimedes' principle 10
Arnfield's escapement 29–30,
  141
atomic clock 11, 121

Bain, Alexander (1810–1877) 84
balance spring 11, 82
barometric compensator 102–4
barometric error 10, 103, 112,
  116
Bateman, Douglas A. 12, 124–5,
  134, 135, 137
Beckett, Sir Edmund, Bt. 159 n.
Big Ben 27, 28, 31, 107
bimetallic strip 116–17

bob 12, 14, 16, 104
Bond's escapement 142–3
Bond, William 139, **140**
Boucheron, Pierre H. (d. 1994)
  135–7
Brain, Peter 60, 149
British Horological Institute 41,
  88, 105
British Museum 51
Brocklesbury Park 40
Brocot, Achille (1817–1878)
  77–80
Buckingham Palace 43
Bulova 8
Burgess, Martin 41, 49, 108
Bush, Vannevar 133

centre of oscillation 102, 116,
  150–1
Chandler clock CM4: 110–16
chaos 135, 137
chime, passing 5
chronometer 132
circular error 36, 41, 77–80,
  133, 148
Clement, William (active
  1671–1694) 21
Coman, William Frederick 1

compensation
  barometric 102–4
  temperature 15, **60**, 110,
  116–17
compliance 9
composer 40
Congreve, Sir William
  (1772–1828) 8, 38, 42–5
conical pendulum **140**, 141
Copley Medal 42
count wheel 32, 33–4, 43–4, **58**,
  58–9
Crane, Aaron Dodd
  (1804–1860) 61
crank escapement 4–5, 28
crown wheel 3
crutch 23
Cumming, Alexander
  (1732–1814) 26

dead-beat escapement 21, 24,
  73
Denison, Edmund Beckett
  159 n.
Dent 83
detachment 42–5, 82, 88, 139
detent 44, 46–8, 97
dominion 11, 41

INDEX

double-beat action 21
drifting rate 107–10
driving force 11, 48, 51, 56, 68, 73
drop 24, 42, 43

earthquakes 137
Earth tides 136–7
elasticity 9
electromagnet 30, 34, 84, 95, 143
endless chain 51, *52*
error
  adopted 109
  barometric 10, 103, 112, 116
  circular 36, 41, 77–80, 133, 148
  escapement 25, 64, 69, 72–7, 79–80
  indicated 121
  random 126
  residual *105*, *110*, 114
  temperature 110, 116–17
escapement error, *see* error
escapements 19–21
  anchor 21–3, **22**
  Brocot 77–80
  Congreve extreme detached 42–5
  constant force 28–9, 35
  dead-beat 21, 24, 73
  double-beat 21
  Froment 28, 30–31
  Goodrich 4, 28
  Graham dead-beat 24, 73
  Harrison grasshopper 38–42, 46, 99
  Kelvin 144–9
  Le Roy 50–5
  lever 82–3
  Nicholson 26
  recoil 20–1, 73, 74–5
  Riefler 20, 84
  verge 3, 25, 64
  Woodward 46–9, **58**, 58–9
  see also gravity escapements, detachment
exponential decay 17–18

flicker floor 124
flicker noise 124–32
flotation 10
fly 28, 98–9
foliot 3, 10
forced vibration 70
Fourier analysis 72–3, 136–7
free pendulum 82–3
  Kelvin **145, 147**, 144–9
  Shortt 84–8, **85**, *86*
  W4: 92–3
  W5: **ii**, 95–7, *99*, *101*
frequency modulation 119
friction drive 56–7, 146
friction rest 24
Froment 25, 28, 30–1
fusee 25

Galileo Galilei (1564–1642) 3, 10, 11
gearing 28, 50, 61
gearless clock 55–60, **58, 60**
George III (1738–1820), King of England 42
Goodrich, Simon 4
Gould, Rupert T. (1890–1948) 132
Graham, George (1675–1751) 24
grasshopper escapement
  Harrison's 38–42
  intermittent 48, **58**, 58–9
gravity escapements 25–6, 139
  Arnfield 29–30, 141
  Bond 142–3
  Froment 30–1
  Grimthorpe 27, 28, 139–40
  Hope-Jones 33–5
  Nicholson 21, 26
  Shortt 83–91
  Thwaites and Reed 27
  Woodward 96–101, *99*, **100**
gravity fluctuations 136–8
Greenwich, Royal Observatory 25, 83, 84, 91
Grimthorpe, Lord (1816–1905) 27, 28, 31, 44, 116, 139
Grollier de Servière 42
Gurney clock 41

hairspring 82, *83*
Hardy, William (active 1800–1830) 115
Harrison, John (1693–1776)
  dominion of pendulum 11, 41
  escapement error 64
  grasshopper escapement 38–42, 46, 99
  gridiron pendulum 15
  maintaining power 48
Hartnup, John 139, 143
Harvard Collection 140 n.
Hipp, Dr Mätthaus (*c.* 1813–1893) 21
hit-or-miss synchronizer 90, 99–101
Hope-Jones, Frank (1867–1950) 33–5, 83, 88, 93, 95
Huber, Johann Jakob 26
Huygens, Christiaan (1629–1695) 3, 41, 51
hysteresis 103

impulse 19–21
inertia 9
  of air 10
  mechanical 9–10
  thermal 16
invar 13, 16
isochronism, lack of 9, 11, 36

Jackson, John Early 133
Jeffreys, Sir Harold 138
Jenkin, Professor Fleeming 147
Jones, Charles Brandram 60

Kater, Captain Henry (1777–1835) 29
Kelvin, Lord (1824–1907) 144–9
King's College, Cambridge 21
Knibb, Joseph (1640–1712) 22, **22**

least squares analysis 113–14
Lepaute, Jean–André (1720–1789) 43, 50–5
Leroy, MM. 84, 91
Le Roy, Pierre (1717–1785) 50–5, **53**

INDEX

lever escapement 82–3
Liverpool Museum 84, **86**, 139, **140**
Louis XV (1710–1774), King of France 50

maintaining power 48, 57–8, **58**
Marcoolyn, Henry 116–17
mass and weight 9–10
Meccano 4–5
momentum 19
motion work 50–1, 55, 59–62
Moxon, J. (c. 1820) 43
Mudge, Thomas (1715–1794) 26

National Bureau of Standards, U. S. A. 7, 122
National Museum of Scotland 115
Newton, Sir Isaac (1642–1727) 13, 63
Nicholson, William (1753–1815) 21, 26
noise 6–7
  pink (flicker noise) 124–32
  red (random walk) 120
  white 118

one-wheel clock 51–5, **53**

pallet 2, 23
pendulum 3
  barometric compensator 102–4
  bob 12, 14, 16, 104
  dominion over clock 11, 41
  gridiron 15
  interaction 92–3, 148
  Riefler 110
  rod 13
  support 12–13, 106
  suspension 14–15, 23
  temperature compensator 15, **60**, 110, 116–17
  theory 9–10
  vacuum working 84, 91
  see also free pendulum, slave pendulum
phase 66–71, 74–6, 80–1, 100–1, 134

Q 11–12
  definitions 17, 68
  in escapement theory 64, 68–9
  measurements 11–12, 17–18, 101
  role 41, 133–5
  in vacuum 91, 135
quartz crystal oscillator 8, 9, 11, 137

radar 6
rainfall 137
random walk 120–1, 126
rate 107–10, 118–19, 121
Rawlings, Arthur L. (c. 1881–1959) 91, 102, 119, 135
reactive force 68
recoil escapement 20–1, 73, 74–5
Redfern, John 115
regulator clock 50, 115, 139
relay 141, 143
remontoire 41, 97–8
resistive force 12, 67–8, 135
resonance 8–11
restoring force 9, 67
Riefler 20, 84, 110
rolling ball 8, 43
Royal Observatory, Greenwich 25, 83, 84, 91
Royal Society (of London) 42, 63
Rugby time signal 118, 122

sample time 123
Sampson, Professor R. A. 108–9
Selticon® (also Secticon) 143
Shannon, Claude E. 6
Shepherd, Charles (c. 1830–1905) 25
Shortt clock 85, **86**
  development 88–9
  Earth tide detection 136–7
  impulse 28, 141
  interaction puzzle 93
  mechanism 84–8
  rate 107–10, 119–23
  stability *124*, *125*
  success of 91
  see also timekeeping charts
Shortt, William Hamilton (c.1881–1971) 35, 83, 108
simple harmonic motion 36
slave pendulum
  unwanted interaction 92–3
  Kelvin's 148
  overloading 94
  Shortt **86**, 89–90
  W5: 96, 100–1
Smith, Humphry 118 n.
stability and instability 123–6
strike, passing 5
suspension 14–15, 23
synchronizer 90, 99–101
Synchronome 33–5, 37, 83, **86**, 89, 118 n.

tangent rule 69
temperature compensation 15, **60**, 110, 116–7
thermal inertia 16
Thomson, Professor Sir William, *see* Kelvin
Thwaites and Reed 27, 140
tides, Earth 136–7
timekeeping charts
  CM4: *111*, *115*
  SH4: *109*, *110*
  SH13, SH41: *122*
  W5: *105*, *106*, *120*
time signal, Rugby 118 n.
Tompion, Thomas (1638–1713) 24
Towneley, Richard (d. 1707) 24
tuning fork 8
turret clock 28, 40, 62

UMAKA clock 1–3, 23
US Naval Observatory 125, 136

vacuum 84, 91
verge 3, 25, 64

165

INDEX

vernier principle 93, 118 n.
volcanoes 137
Vulliamy, Benjamin Lewis
    (1780–1854) 43

W1: 32
W2: 33, 35, 37
W3: 46–9, 59

W4: 92–4
W5: **ii**, 28, 95–106, **100**, 143
Wadham College, Oxford 21–2, 22
Wallman, Professor Henry
    (d. 1992) 132, 137–8
watch
  balance spring 11
  detached lever 82–3

duplex escapement 55
tuning fork 8
verge 25, 64
weight and mass 9–10
Worth Matravers 6
Wren, Sir Christopher
    (1632–1723) 22
Wright, John F. 141

IMAGES OF WAR
STAR-SPANGLED SPITFIRES

A PHOTOGRAPHIC RECORD OF SPITFIRES FLOWN BY AMERICAN UNITS

IMAGES OF WAR
STAR-SPANGLED SPITFIRES

A PHOTOGRAPHIC RECORD OF SPITFIRES FLOWN BY AMERICAN UNITS

TONY HOLMES

Pen & Sword
AVIATION

First published in Great Britain in 2017 by
PEN & SWORD AVIATION
an imprint of
Pen & Sword Books Ltd,
47 Church Street, Barnsley,
South Yorkshire.
S70 2AS

Copyright © Tony Holmes 2017

ISBN 978-1-47388-923-1

The right of Tony Holmes to be identified as Author of
this Work has been asserted by him in accordance with the
Copyright, Designs and Patents Act 1988.

A CIP catalogue record for this book is available
from the British Library

All rights reserved. No part of this book may be reproduced or transmitted
in any form or by any means, electronic or mechanical including photocopying,
recording or by any information storage and retrieval system,
without permission from the Publisher in writing.

Typeset by Mac Style Ltd, Bridlington, East Yorkshire
Printed and bound in India by Replika Press Pvt Ltd.

Pen & Sword Books Ltd incorporates the imprints of
Pen & Sword Aviation, Pen & Sword Family History, Pen & Sword Maritime,
Pen & Sword Military, Pen & Sword Discovery, Wharncliffe Local History,
Wharncliffe True Crime, Wharncliffe Transport, Pen and Sword Select,
Pen and Sword Military Classics

For a complete list of Pen & Sword titles please contact:
Pen & Sword Books limited
47 Church Street, Barnsley, South Yorkshire, S70 2AS, England.
E-mail: enquiries@pen-and-sword.co.uk
Website: www.pen-and-sword.co.uk

Contents

Acknowledgements vi
Introduction vii

Chapter 1: Early Operations in the ETO 1

Chapter 2: Combat in the MTO 34

Chapter 3: Training and Photo-Reconnaissance Units 79

Chapter 4: Spitfires in Glorious Colour 87

Acknowledgements

The author wishes to thank the following individuals (some of whom, sadly, are no longer with us) and organizations for the provision of photographs and information included within this volume:

Peter Arnold, Norman Franks, the late Roger Freeman, Peter Green, William Hess, Philip Kaplan, Paul Ludwig, Dick Martin, Wojtek Matusiak, the late Bruce Robertson, Andy Saunders, the late Jerry Scutts, Sam Sox and Andrew Thomas.

Introduction

As a follow-on to my *Images of War – American Eagles* volume of 2015, this book focuses on the iconic Spitfire marked with the equally distinctive USAAF star (and later bars). Three fighter groups, each consisting of three squadrons, would see brief combat with the Supermarine fighter in the European Theatre of Operations (ETO) during the late summer and autumn of 1942. Equipped with Spitfire VBs (the most-produced mark), the 4th, 31st and 52nd Fighter Groups would enjoy modest success on the Channel Front prior to the latter two units being sent to support the American-led invasion of North Africa – codenamed Operation *Torch* – in November 1942. The 4th FG, manned in the main by pilots who had previously seen combat with the RAF's trio of 'Eagle' squadrons prior to them being transferred to USAAF control in late September 1942, continued to fly the Spitfire VB in the ETO until it switched to the P-47 Thunderbolt from March 1943.

By then, the 31st and 52nd FGs had become well and truly embroiled in the Mediterranean Theatre of Operations (MTO), flying tropicalized Spitfire VBs and hard-hitting 'quad cannon' VCs against German and Italian fighters and bombers in the war-torn skies over Tunisia as the Allies slowly got the better of the *Afrika Korps*. Assigned to the Twelfth Air Force and flying alongside P-38 Lightning, P-39 Airacobra and P-40 Warhawk fighters that equipped other USAAF fighter groups in-theatre, the two Spitfire units more than held their own in traditional fighter missions and in the demanding fighter-bomber role. Supporting troops on the ground grew in importance once all Axis forces had been defeated in North Africa and the Allies turned their attention to Italy. From mid-1943 the two groups started to replace their war-weary Spitfire Vs with vastly superior Mk IXs, even better Mk VIIIs arriving by the end of the year. Making the most of their mount's outstanding abilities as a fighter, some twenty-two USAAF pilots had claimed five or more victories to 'make ace' on the Spitfire in the MTO by the time the final examples were replaced by P-51B/C Mustangs in the early spring of 1944.

In the ETO, Spitfires had equipped the tactical reconnaissance (TAC-R) optimized 67th Reconnaissance Group (RG) following its arrival in England in the autumn of 1942, many of its aeroplanes being cast-offs from the 31st and 52nd FGs after the units headed for North Africa minus their Mk VBs. Although these machines lacked cameras, they served as ideal mounts for the intensive training undertaken by 67th RG pilots as they learned how to observe enemy targets and strafe them effectively. In late 1943 the group transferred from the strategic Eighth Air Force to the tactical Ninth Air Force, and in January 1944 the 67th RG was issued with

TAC-R F-6 Mustangs. Having never fired a shot in anger with the group, the last war-weary examples of its Spitfire VBs were retired by the time the 67th moved to France in July 1944.

The final 'star-spangled' Spitfires in the frontline ranks of the Eighth Air Force were the high-flying, and unarmed, PR XI photo-reconnaissance aircraft supplied to the 7th Photographic Reconnaissance Group to supplant its F-5 Lightnings from November 1943. Ranging as far into Germany as Berlin, the 'PR blue' Spitfires provided critical target imagery – both pre- and post-strike – for the 'Mighty Eighth's' heavy bombardment groups through to April 1945. Flying exclusively with the 14th Photographic Reconnaissance Squadron from January 1944, these aircraft performed myriad missions alongside Lightnings and, eventually, P-51 Mustangs.

Only a handful of British combat aircraft wore the 'stars and bars' of the USAAF in the Second World War, with the Beaufighter, Mosquito and Spitfire being the key types to see action with American crews in American squadrons. The Spitfire was, by some margin, the most widely used of the three, and the 'Yanks' that flew it in combat rated the fighter very highly.

Tony Holmes
Sevenoaks, Kent
August 2016

Chapter One

Early Operations in the ETO

In order to gain operational experience of combat in the European Theatre of Operations (ETO), three senior officers from the 31st Fighter Group (FG) initially flew under the auspices of No 412 Sqn of the Royal Canadian Air Force. Unfortunately, during their first sortie – a 'Rodeo' (fighter sweep) to the Luftwaffe fighter airfield at Abbeville – on 26 July 1942, the group's Executive Officer, Lieutenant Colonel Albert Clark, flying Spitfire VB BL964/VZ-G, was jumped by Fw 190s from *Jagdgeschwader* (JG) 26 and shot down. He was quickly taken prisoner, his captors being baffled by his USAAF uniform and high rank. Clark had become the first operational USAAF fighter casualty in Europe.

Few photographs exist of the 52nd FG's Spitfires in the ETO, despite its three squadrons being fully equipped with the British fighter from July 1942. These Mk VBs, assigned to the group's 2nd Fighter Squadron (FS), are seen basking in the summer sun at either Eglinton, in Northern Ireland, or Atcham, in Shropshire. Note the unusual camouflage patterns on the fighters' tails where the RAF fin flash has been painted out by hand, while QP-Z in the foreground also has a lighter patch on its upper wing surface obscuring the roundel.

From this angle, the 2nd FS Spitfire VB to right appears to still have an RAF roundel beneath its port wing. The USAAF national insignia was usually applied beneath the starboard wing. The aircraft third from the right has faded white vertical stripes on its nose, denoting its previous assignment to an RAF Fighter Command unit that was scheduled to participate in the cancelled July 1942 raid on Dieppe, codenamed Operation *Rutter*. These distinctive markings had been removed by the time the raid took place on 19 August that same year.

For some reason this Spitfire VB of the 308th FS/31st FG, seen at Kenley, in Surrey, in August 1942, is fitted with an early pattern 'blunt' de Havilland propeller spinner. Most Mk Vs featured the more common, longer, Rotol spinner. The aeroplane was routinely flown during this period by future ace Captain Frank Hill, although it is not known if this is the Spitfire he was flying when he made his first claim – for an Fw 190 probably destroyed – over Dieppe on 19 August 1942.

Lieutenant Edward Dalrymple, who would later achieve two victories, is seen at the controls of HL-C during a convoy patrol over the Channel from Kenley in August 1942. This is the aeroplane that was also flown by Captain Frank Hill. The fighter's de Havilland spinner is clearly visible in this view. The Rotol spinner was usually seen on Mk Vs, with the 'blunt' de Havilland unit being almost a 'standard fit' for early marks of Spitfires. This would suggest that HL-C may have been a converted Mk I or II.

Major Harrison Thyng strikes a typical fighter pilot's pose while undertaking his conversion on to the Spitfire in late June 1942 at Atcham. Note that his aircraft has yet to have its RAF roundel replaced with a USAAF star. Having joined the army as an infantryman in 1939, Thyng had subsequently gained his wings and seen service with the 1st Pursuit Group (PG) prior to joining the newly formed 31st PG in October 1940. The first CO of the 309th Pursuit Squadron (PS), which was created in January 1942 to replace the 41st PS after the latter was sent to Australia to fight the Japanese as part of the 35th PG, Thyng led the squadron to the UK. He duly claimed the 309th's first combat successes, taking his score to five victories – and thereby 'making ace' – by the time he completed his tour in mid 1943. Thyng saw further action as CO of the P-47N Thunderbolt-equipped 413th FG in the Pacific in 1944–45. Remaining in the regular air force post-war, and promoted to colonel, Harry Thyng went on to raise his tally of victories to ten while flying F-86A/E Sabres as CO of the 4th Fighter Interceptor Wing in Korea in 1951–52.

Subtly marked with a Star of David, this Spitfire VB was the personal mount of First Lieutenant 'Buck' Inghram, who can be seen sat in its cockpit at Kenley in early August 1942. Assigned to the 31st FG's 308th FS, Inghram was the first of seven American pilots from the group to be shot down over Dieppe on 19 August while supporting Operation *Jubilee* – the ill-fated Dieppe raid. Flying as part of the Kenley Wing on an early morning low-level sweep over the invasion beaches, he and his squadron were 'bounced' by between twenty and thirty Fw 190s from II./JG 26. Inghram's Spitfire was almost certainly shot down by 4. *Staffel*'s Oberfeldwebel Wilhelm Philipp, the American being his nineteenth victory out of an eventual total of eighty-one. Parachuting down into the Channel, Inghram was eventually captured after drifting ashore in his dinghy.

Spitfire VB BM587 of the 309th FS/31st FG is serviced at Westhampnett, in West Sussex, in early August 1942. Delivered new to the RAF on 7 May 1942, the Merlin 45-powered fighter was passed on to the 309th at High Ercall, in Shropshire, on 20 June. The aeroplane was returned to RAF service on 12 September and served with a handful of frontline fighter units until it was written off when the pilot overshot his landing at Northolt, in Middlesex, on 2 January 1945.

Two members of Major General 'Monk' Hunter's staff pose in front of the 31st FG's very first Spitfire during a visit by the VIII Fighter Command boss to Biggin Hill, in Kent, during the afternoon of 13 August 1942. The individual on the right is Major J. Francis Taylor, while his colleague remains unidentified. Accompanying Hunter to Kent's premier station was Major General Carl Spaatz, Commander of the Eighth Air Force. Both men were keen to check on the rate of progress made by the 307th FS since its posting to Biggin Hill on 1 August, and to gauge the unit's state of preparedness for action.

Lieutenant R. Wooten of the 307th FS/31st FG is helped on with his parachute prior to climbing into the cockpit of his Spitfire VB EN851 at Merston, in West Sussex, in late August 1942. The aircraft was the first of three presentation Spitfires paid for by donor Mr H.L. Woodhouse of Lima, Peru, hence the *Lima Challenger* titling forward of the cockpit. Following its service with the USAAF (which included employment as a training aircraft with the 107th Observation Squadron (OS) following the 31st FG's transfer to the MTO), EN851 was transferred to the Fleet Air Arm in February 1943 and converted into Seafire IB NX952 by Cunliffe–Owen Aircraft Ltd. The fighter was eventually written off in an accident on 16 September 1944 while serving with Fleet Air Arm training unit 761 Naval Air Squadron.

On 22 September 1942 the press visited the 309th FS at Westhampnett (now Goodwood airfield), where the unit entertained them by conducting the usual 'stunts' associated with such occasions – mock scrambles, formation take-offs, airfield 'beat ups' and posed individual and group shots. With the scramble bell ringing in their ears, a clutch of pilots run from the dispersal hut for their Spitfires. The aircraft parked on the finger of tarmacadam behind them is Mk VB EP179, which was assigned to Major Harrison Thyng, CO of the 309th – he is the pilot furthest to left in this photograph. Having been credited with damaging two Fw 190s since arriving at the West Sussex airfield on 4 August, Thyng claimed a Ju 88 probably destroyed in EP179 on the very day this photograph was taken. EP179 had been delivered new to No 71 'Eagle' Sqn on 20 July 1942, the fighter being passed on to the 309th FS two months later. It returned to the unit (which had since become the 334th FS of the 4th FG when transferred to the USAAF in September 1942) on 1 October and was eventually assigned to the 15th OS of the 67th Observation Group (OG). EP179 was subsequently supplied to the *Armée de l'Air* in January 1945 and served with *Groupe de Combat* II/18.

Prior to 'scrambling' for their aircraft at Westhampnett on 22 September 1942, pilots from the 309th FS conducted a mock briefing below a rather limp 'Old Glory'. The individual pointing at the map is Major Harrison Thyng.

Staff Sergeant Olin M. Battles of Hartselle, Alabama, carries out a sixty-hour inspection on a 309th FS Spitfire VB in one of the temporary Miskin steel blister hangars erected at Westhampnett. Like most Supermarine fighters assigned to the 31st FG in Britain, this aircraft has been personalized through the addition of a nickname below the windscreen.

On 24 August 1942 the 308th FS joined the 309th at Westhampnett, the unit flying in from Kenley so as to allow the 31st FG to operate more as a group – the 307th FS had arrived at the airfield that same day, only to be sent to nearby Merston twenty-four hours later. This photograph of the 308th's 'B' Flight was taken at Westhampnett during September. These pilots are, from left to right, Charles Van Reed, Mathew Mosby, Adrian Davis, E.G. Johnson, Derwood Smith, Westley Ballard, John Ramer and Frank Hill. The latter would later become the 31st's top-scoring Spitfire ace with seven victories, two probables and five damaged (all bar a single probable being claimed in the MTO). He also led both the 308th and 309th FSs prior to assuming command of the 31st FG in mid-July 1943.

A 308th FS Spitfire VB is serviced between flights at Westhampnett on 3 September 1942. One groundcrewman is refilling the fighter's 48-gallon main fuselage fuel tank from a nearby bowser, another is giving the blown canopy a good polish and a third individual is checking the frequency settings for the aircraft's transmitter/receiver, installed in its own compartment below the aerial mast. The constant repetition of this procedure soon revealed the bare metal of the Spitfire's wing root, which has been inexplicably touched up with insignia white paint in this particular instance.

First Lieutenant R.F. Sargent (right) enjoys a cigarette while helping First Lieutenant E.S. Schofield with his parachute straps between flights at an unidentified airfield (possibly Debden, in Essex) in early October 1942. Schofield is wearing an RAF issue type C-2 seat pack parachute over his standard USAAF A-4 summer suit. Sargent, however, appears to have on RAF battledress, over which he is wearing a bulky Thermally Insulated Flying Jacket, better known as an Irvin jacket. His footwear is also of British origin in the form of 1941 pattern flying boots. The well-weathered Spitfire bears both the MX codes and squadron badge of the 307th FS.

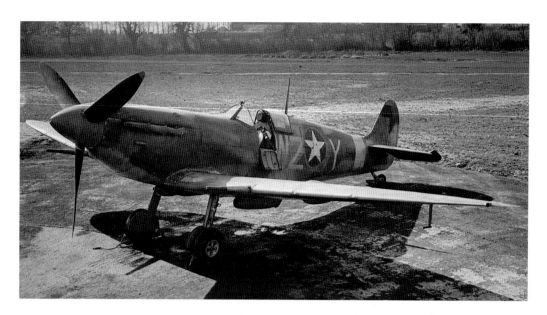

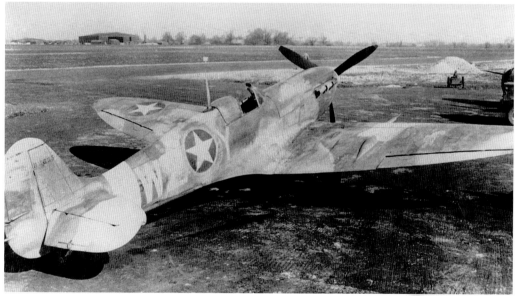

Spitfire VB BM635 was among the batch of new aircraft delivered to High Ercall, in Shropshire, on 21 June 1942 for use by the 31st FG upon the group's arrival in the UK. Flown by future ace Second Lieutenant Dale Shafer, among numerous other pilots, during the summer, it was slightly damaged by him in a flying accident on 20 July. The fighter remained with the 309th FS until the unit was withdrawn from operations in October prior to it being sent to North Africa with the rest of the 31st FG. As these two photographs clearly reveal, BM635 had been well used during its time with the Americans, its day fighter scheme looking decidedly the worst for wear by the late autumn of 1942. Like numerous ex-31st FG Spitfires, this aircraft was supplied to the 67th OG in October 1942, the aeroplane serving with the 109th OS at Membury, in Berkshire. BM635 was returned to the RAF in November 1943 and eventually struck off charge in February 1945.

Debden was the operational home of the 4th FG from its inception in September 1942 through to war's end. This aerial view of the fighter station was taken in late 1944 – long after the 'star-spangled' Spitfires had been replaced by US-built machines – although the main hangars and general layout had changed very little since the RAF's trio of US-manned 'Eagle' squadrons had been transferred to the USAAF to form the 4th FG on 29 September 1942. Built between 1935 and 1939, Debden was a sector station for Fighter Command's No 11 Group during the Battle of Britain and became home for the Debden Wing until it was turned over to the USAAF for use by the Eighth Air Force. This shot clearly reveals how the airfield was dominated by three 152ft span C-type hangars, with two intersecting runways of 1,600 and 1,300 yards visible in the distance. Aside from the aircraft in front of the hangars, other machines can be seen dispersed along the perimeter track that runs around the edge of the airfield – particularly in front of Abbotts Farm, which continues up to the runway's point of intersection.

When the 'Eagle' squadrons joined the Eighth Air Force in September 1942, 22-year-old Chesley Peterson was transferred from the RAF to the USAAF as a lieutenant colonel and became XO of the 4th FG. By then one of the most combat-experienced American pilots in the ETO, Peterson was one of a number of pilots from the 4th that participated in a press photographic session staged at Debden in late March 1943. The group's Spitfires also featured prominently on the day, despite them having by then been taken off offensive operations while the 4th transitioned on to the P-47 Thunderbolt. The USAAF was keen to create the impression that the Supermarine fighter was still in active use, thus tricking enemy intelligence into believing that American Spitfires remained a part of the frontline force.

Future 4th FG CO Lieutenant Colonel Don Blakeslee is flanked by 334th FS pilots First Lieutenant Spiros 'Steve' Pissanos and Captain Vernon Boehle. They are posing in front of a 336th FS Spitfire VB that has been marked up with the unit's boxing eagle emblem, previously worn by aircraft of No 71 'Eagle' Sqn – all three pilots had transferred from the RAF to the USAAF when the 'Eagle' squadrons became part of VIII Fighter Command. Both Pissanos and Blakeslee would subsequently become aces, while Boehle was posted to the Ninth Air Force after completing his tour with the 4th FG in November 1943.

This heavily censored photograph shows Spitfire VB EN783 at Debden (the control tower has been officially removed from the background) in late 1942. Wearing XR codes of the 334th FS, this aeroplane had been assigned to No 71 Sqn when the trio of American-manned units were transferred to the USAAF and serving with the 4th FG. A much-travelled aircraft, it had originally been delivered new to No 610 Sqn on 21 May 1942 and was then passed on to the 31st FG's 308th FS on 15 July. EN783 remained within the group when it was issued to the 309th FS on 24 August, and the unit in turn sent it to No 71 Sqn nineteen days later. Re-engined with a Merlin 46 and mechanically upgraded by Vickers-Supermarine once declared surplus to requirements by the 4th FG in March 1943, the aeroplane saw fleeting frontline service with Nos 66 and 340 Sqns eight months later. Damaged in combat on 7 December, EN783 was subsequently repaired and sent to No 1 Tactical Exercise Unit (TEU) on 16 June 1944, followed by No 57 Operational Training Unit (OTU) on 12 September. The veteran fighter was finally written off on 2 May 1945 when engine failure caused its pilot to crash-land in a field near Eshott, in Northumberland.

An unidentified pilot poses on the wing of a 334th FS Spitfire VB at Debden while his equally anonymous squadronmate extricates himself from the fighter's cockpit. Note how exhaust-streaked and chipped the paintwork is, and that both the undercarriage legs and the radiator fairing are caked in dried mud.

Second Lieutenant Don Gentile poses in front of his uniquely marked Spitfire VB BL255 at Debden soon after joining the USAAF in September 1942. A member of the 336th FS, he would go on to become the unit's top-ranking ace with 21.833 kills by the time he returned to the USA in late April 1944. BL255 was the only aircraft that Gentile christened *Buckeye-Don*, its P-47D replacement being called *Donnie Boy* and the P-51B that in turn followed in March 1944 bearing the name *Shangri-La*. All three aircraft were, however, adorned with the boxing eagle motif that eventually became the emblem of the 336th FS. BL255 features two victory markings, which denote Gentile's Ju 88 and Fw 190 claims on 19 August 1942 just east of Dieppe. He was serving with No 133 'Eagle' Sqn (which duly became the 336th FS in late September 1942) at the time. Prior to enjoying a long spell of service firstly with No 133 Sqn and then the 336th FS, BL255 had flown with No 611 Sqn in early 1942. Issued to No 610 Sqn after being discarded by the USAAF in the spring of 1943, the fighter was eventually passed on to No 118 Sqn in the Orkneys in late May 1944 and then returned to No 611 Sqn in early October. Just days later BL255 was relegated to training duties with No 61 OTU and was struck off charge on 22 May 1945.

Pulled off readiness by visiting press photographers and asked to stand in front of the 336th FS's scoreboard, painted on a wall in the squadron's operations building at Debden, Second Lieutenant Don Gentile smiles for the camera. The many crosses that adorn the board denote kills, probables and damaged claims credited to the unit during its time with No 133 Sqn within RAF Fighter Command. The original artwork for the Air Ministry-approved squadron crest can also be seen hanging in a frame above the eagle.

Six-victory ace Second Lieutenant Roy Evans opened his account while flying Spitfires with the 335th FS, downing a Fieseler Fi 156 Storch near Furnes during a sortie over France on 21 November 1942. Another 'Eagle' squadron pilot (he flew with No 121 Sqn), Evans eventually became deputy CO of the 359th FG and ended the war as a PoW after being shot down in a P-51D Mustang on 14 February 1945.

Opposite: Ex-No 121 'Eagle' Sqn pilot Captain Don Willis was made Operations Officer of the 335th FS following the former unit's absorption into the USAAF. Having served in the RAF since late 1941, Willis was a vastly experienced pilot who wore no fewer than four sets of wings on his uniform – Finnish Air Force, Royal Norwegian Air Force (RNAF), RAF and USAAF. Almost certainly the first US-based volunteer pilot of the Second World War, Willis had trained with the Finns during the Russo–Finnish War of late 1939, then joined the RNAF when Germany invaded in April 1940. Escaping to the UK shortly before Norway fell, Willis eventually made it into the RAF and then to No 121 Sqn. Tour-expired in 1943, he returned to action the following year but was shot down in a P-38 Lightning and made a PoW until war's end. Note the white cross painted on to Willis's helmet, which was added to make the wearer more visible should he be forced to bail out over the sea. It is highly likely the Spitfire VB that Willis is sat in is BM590, which had served with No 121 Sqn as AV-R from new (it was delivered to the RAF on 24 April 1942). Passed on to the 13th Photographic Reconnaissance Squadron (PRS) of the 7th Photographic Reconnaissance Group (PRG) at Mount Farm, in Oxfordshire, in August 1943, the aeroplane was written off on 7 October that same year. The fighter, with Lieutenant V.N. Luber at the controls, had suffered engine failure while taking off from the airfield.

Former schoolteacher Major Oscar Coen (right) had a colourful career with the 'Eagles', having been shot down and evaded capture to return to duty. He became a squadron commander on transferring to the USAAF, leading both the 334th and 336th FSs and, later in the war, the 356th FG. By war's end Coen had claimed four victories, 1½ probables and three damaged. The bulk of his successes came in the Spitfire VB, including two victories (both Fw 190s) scored with the 4th FG.

Although its serial is hidden by the fighter band, this Spitfire VB is almost certainly EN793, which was the mount of 336th FS CO and eight-victory ace Major Carroll 'Red' McColpin up until his return to the USA on 29 November 1942. Issued new to No 137 Sqn on 15 June 1942, the fighter was sent to No 121 Sqn just days later and then transferred to No 71 Sqn on 28 July. EN793 was transferred to No 306 'Polish' Sqn on 20 August, before finally reaching No 133 Sqn on 29 September. The aircraft suffered an undercarriage collapse on landing back at Debden on 22 January 1943 after it received battle damage during the 4th FG's final large-scale engagement with the Spitfire. Soon repaired, EN793 was re-engined with a Merlin 46 following the 4th FG's re-equipment with P-47s. Issued to No 317 'Polish' Sqn on 9 September, the fighter was sent to No 312 'Czech' Sqn the following month. Here it remained until 27 February 1944, when the Spitfire was passed on to No 443 'Canadian' Sqn. Relegated to No 1 TEU on 26 April, EN793 was finally sent to No 61 OTU on 25 June and struck off charge just six days after VE Day.

Spitfire VBs BL255 MD-T and MD-V (serial unknown) 'beat up' Debden for the benefit of the visiting press corps. The pilot of the former is almost certainly Second Lieutenant Don Gentile, who had a penchant for such flying.

Undoubtedly one of these individuals was at the controls of MD-V during the 'beat-up' photograph seen previously in this chapter. Sadly, their identities remain unknown. As with all USAAF Spitfires in the ETO, this example looks well used.

Armourers work on the port cannon of Spitfire VB BL766 in front of the 336th FS's perimeter HQ shack at Debden in late March 1943. This aircraft was often flown by Major Don Blakeslee during his time as squadron CO. The 336th was the last unit to transition to the Thunderbolt within the 4th FG, flying its final operation mission with the Spitfire on 10 April 1943.

Spitfire VB EN853 was the personal mount of the 335th FS's first CO, and ace, Major Jim Daley, until he returned to the USA tour-expired on 22 November 1942. His final claim in the ETO was almost certainly achieved in this aircraft, Daley being credited with damaging an Fw 190 east of Calais while leading his unit on an escort mission for a diversionary bombing force from the Eighth Air Force on 2 October 1942. Yet another aircraft assigned to an 'Eagle' squadron at the time the units were transferred to USAAF control, EN853 had initially seen frontline service with No 401 'Canadian' Sqn from 4 June 1942 through to 5 August, when it was passed on to No 121 'Eagle' Sqn. The fighter was subsequently shot down by flak while escorting RAF Bostons sent to bomb the airfield at Saint-Omer on the afternoon of 22 January 1943. Its pilot, Second Lieutenant Chester Grimm, was seen to bail out of the stricken Spitfire off Dunkirk, but his body was never found.

Members of the 4th FG's maintenance section carry out what must have been one of the very last engine changes performed on a Spitfire at Debden. Photographed inside a large C-type hangar at the base in early April 1943, this Mk VB wears the MD codes of the 336th FS. By the end of that month only a solitary example of Supermarine's superlative fighter remained at Debden.

Lieutenant 'Bill' Chick of the 336th FS poses in the cockpit of the Spitfire VB assigned to XO of the 4th FG, Lieutenant Colonel Chesley 'Pete' Peterson, in January 1943. Like his RAF equivalents in Fighter Command, Peterson identified his aircraft (almost certainly BL449) with his initials CG-P. The first American 'Eagle' to command a squadron, Peterson achieved six of his seven victories in the Spitfire when with No 71 Sqn. He was transferred to the USAAF as a lieutenant colonel and observed the RAF practice of wing leaders having their initials applied to their assigned aircraft in place of unit code letters. This Spitfire is believed to have been the very first USAAF fighter to have worn a personal code. If CG-P was indeed BL449, then it was transferred back to the RAF in early 1943 and eventually passed on to the Portuguese Air Force in December of that same year.

Spitfire VB BM309 of the 335th FS appears to be suffering from some kind of oil or glycol leak judging by the large puddle of liquid forming beneath its engine. Photographed within one of the many blast pens scattered around the perimeter track at Debden in February 1943, the fighter sits opposite a clipped wing Spitfire VB of the 336th FS. The previous month, on 22 January, BM309 had been used by Second Lieutenant Robert A. Boock to destroy an Fw 190 north-west of Dunkirk in the 4th FG's last big engagement with the Spitfire. Returned to the RAF shortly thereafter, BM309 served with Polish-manned Nos 315 and 303 Sqns and, finally, No 313 'Czech' Sqn between March and November 1943, when it was supplied to the Fleet Air Arm and converted into a Seafire IB by Cunliffe–Owen.

Chapter Two

Combat in the MTO

This was the sight that greeted the pilots of the 31st and 52nd FGs upon their arrival in Gibraltar from Scotland in early November 1942 – brand new Spitfire VBs and VCs complete with unusually proportioned national markings outlined with a yellow ring applied to all Allied aircraft involved in the American-led *Torch* landings in North Africa. Also featuring Vokes tropical dust filters, each of the Spitfires boasted RAF fin flashes, which were rarely seen on aircraft assigned to the Twelfth Air Force. Spitfire VB ER219 in the centre of the photograph arrived in Gibraltar on 1 November 1942 and it served until struck off charge on 30 April 1943, its final fate unknown.

The 31st FG's 308th FS initially saw little action over Algeria after flying in from Gibraltar following the American-led *Torch* landings in North Africa on 8 November 1942. The unit did suffer occasional accidental losses, however, although the damage inflicted on this Spitfire VC, which crash-landed at Maison Blanche on 12 December 1942, appears to be repairable.

The wreckage of Spitfire VC ER488 is spread over the Algerian landscape after it crashed on 17 December 1942. The unit to which this aeroplane was assigned and its pilot at the time of the fighter's demise remain unrecorded.

2nd FS/52nd FG Spitfire VB QP-L (serial number EP???) came to grief possibly at La Sebala, in Tunisia, in 1943. The aeroplane is being inspected by local tribesmen, one of whom appears to being wearing a German paratrooper's helmet.

The first pilot to become an ace flying the Spitfire in USAAF service was Captain Jimmie Peck of the 2nd FS, who claimed his fifth, and last, victory early on 2 January 1943 when he shot down an Fw 190 near Bone. His previous successes had come while serving with the RAF in the defence of Malta during 1942 – he is seen here in British uniform.

Attrition was high within the 31st and 52nd FGs during the early months of the North African campaign due to the harsh meteorological conditions in-theatre and the battle-hardened Axis fighter units encountered over the frontline. This anonymous Spitfire VB from the 52nd FG's 5th FS was crash-landed in northern Tunisia during the spring of 1943. The American flag applied in decal form beneath the cockpit was an additional recognition marking added by USAAF units to their aircraft for the benefit of the Moroccan and Algerian locals. The flag decal was worn on both sides of the fuselage.

Spitfire VC ES306 HL-D of the 308th FS/31st FG sits at Thelepte, in western Tunisia, in March 1943, this airfield having actually been hastily evacuated by the unit the previous month in the face of an Axis offensive. At the time this photograph was taken the squadron was in the midst of covering the US Army's advance towards Tunis, which came under persistent enemy air attack. Thelepte, like most airfields in North Africa, was little more than a patch of flat ground close to the constantly shifting frontline. Note the groundcrew shelter in the right background behind the Spitfire – forward airstrips such as Thelepte were routinely strafed by marauding Axis fighters. Surviving almost a year on operations, this aeroplane was flown regularly by future seven-victory ace Captain Frank Hill during the early spring of 1943. Eventually returned to RAF service, the fighter was assigned to No 249 Sqn and written off in a forced landing after it was struck by flak near Rozge, in Yugoslavia, on 16 December 1943.

CO of the 52nd FG's 4th FS for much of the North African campaign, Major Robert Levine claimed all his victories flying his personal Spitfire VC ER570 WD-Q. The latter is seen here at La Sebala in the spring of 1943, displaying a swastika for Levine's first victory, which he claimed on 8 January. The 4th FS's CO from mid 1942, Robert Levine was promoted to major on 1 January 1943. A respected leader, he scored his own, and his squadron's, first success in early January when he used ER570 to shoot down an Fw 190 – this victory was duly recorded on the nose of the aircraft. Like other Spitfires of the 4th FS at the time, ER570 bore its pilot's name in script, as well as the 4th FS's striking skull insignia, three victory symbols and a hand-painted 'Old Glory' on either side of the fuselage. He was also flying this aircraft on 19 April when, over La Sebala, he shot down a Bf 109. Levine's third, and final, victory came the next day when he bagged another Messerschmitt, probably from II./JG 51. These victories, plus one probable, were all claimed in ER570. In February 1944 Levine became the group commander, while ER570 was later returned to the RAF and eventually scrapped in early 1945.

Future Spitfire ace Captain Arnold 'Vince' Vinson of the 2nd FS/52nd FG achieved his first claims on 2 December 1942 when he damaged a Bf 109 and then shot down an Fw 190 west of Bizerte. He described his first victory as follows in his combat report: 'I gave the enemy aircraft a one-second burst. Strikes were seen as the enemy aircraft weaved through a valley and I closed to 50 yards, striking the engine, and it smoked and stopped operating.' Vinson became an ace on the evening of 24 March 1943, but on 3 April, having just destroyed a Ju 87, he was jumped by a Bf 109 and killed while flying Spitfire VB JK171.

The 2nd FS's Captain Norman McDonald, from North Carolina, attained ace status in spectacular fashion on the evening of 3 April 1943 when he downed three of the thirteen Ju 87 Stukas from III./StG 3 claimed destroyed by American Spitfires south-east of El Guettar, in Tunisia. He was one of the leading USAAF Spitfire pilots, with 7½ of his eventual total of 11½ kills being claimed with the British fighter.

During a large-scale dogfight over Kairouan, in Tunisia, on 9 April 1943, the 52nd FG claimed eleven victories and First Lieutenant Vic Cabas became an ace. Future aces First Lieutenants Sylvan Feld, Fred Ohr and John Aitken and Staff Sergeant James Butler also claimed. One casualty, however, was First Lieutenant Eugene Steinbrenner, whose Spitfire VC ER120 VF-D of the 5th FS crash-landed on a ridge after being struck by flak. The aeroplane was later recovered and stripped of all salvageable parts (a process that is clearly under way in this photograph), before being struck off charge. Behind the fighter is another Spitfire VB from the 5th FS, as well as the engineless fuselage of a B-26 Marauder and the wings of a P-40 Warhawk.

Captain Moss Fletcher of the 4th FS also became an ace during the heavy fighting of April 1943, having claimed his first (shared) victory while flying with the RCAF. During his time with the 4th he was credited with a further four kills in late April. Fletcher's aircraft not only carried his name and score, but also personal nose art. His crew chief, Sergeant Kurka, is seen here on the wing of Fletcher's Spitfire.

A smiling First Lieutenant Jerry 'J.D.' Collinsworth of the 307th FS/31st FG poses for the camera in the North African sun. He claimed four victories during the fighting over Tunisia in the spring of 1943, and subsequently became an ace during the bombardment of Pantelleria in early June.

Spitfire VC ER187 WZ-C of the 309th FS/31st FG was flown by unit CO Major Frank Hill on the afternoon of 6 May 1943 when, during a sweep of the Tunis area, he claimed a Bf 109 shot down and damaged a second Messerschmitt fighter and an Italian Macchi C.202. Earlier that same day he had 'made ace' when he destroyed two more Bf 109s while flying one of the first Spitfire IXs issued to the 31st FG. A designated alert fighter, ER187 is connected to a starter trolley (beneath its starboard wing).

New Yorker Major Frank Hill poses alongside his Spitfire VC ER187 WZ-C, which bears his victory total and the names of both him and his wife. By the time he was promoted to CO of the 31st FG on 15 July 1943 he had claimed his seventh, and last, aerial victory.

Lieutenant Colonel Frank Hill replaced Lieutenant Colonel Fred M. Dean as CO of the 31st FG, the latter having led the group throughout the North African campaign. Dean had claimed his only victory during the Pantelleria operation in June 1943.

Spitfire VC ER256 was the personal aircraft of Lieutenant Colonel Fred Dean, who, as CO of the 31st FG, adopted the RAF practice of carrying his initials on his aircraft rather than squadron codes.

With all Axis forces defeated in North Africa, Spitfire VC ES138 WD-N of the 4th FS/52nd FG sits on a dusty Tunisian airfield during the summer of 1943 prior to the group moving across the Mediterranean to Sicily. Like many other USAAF Spitfire Vs that saw combat in the MTO, the final fate of this aircraft remains unrecorded.

Lieutenant Colonel R.A. Ames, XO of the 31st FG, formates with his wingman over a choppy Mediterranean Sea off the Tunisian coast in Spitfire VC JK226 HL-AA of the 308th FS during the spring of 1943. Following service with the USAAF, this aircraft was supplied to No 32 Sqn and saw combat in Italy, before eventually being handed over to the recently formed Greek Air Force in 1946.

Spitfire VC ES364 WD-F of the 4th FS crash-landed in Tunisia in mid 1943, possibly after operations in support of the landings on Pantelleria. Unusually, this aircraft still has an RAF fin flash – it was possibly an attrition replacement supplied to the 52nd FG following the heavy losses suffered by the USAAF Spitfire units during the final stages of the Tunisia campaign. Duly repaired, ES364 was returned to RAF control towards the end of 1943.

In preparation for the invasion of Sicily, the 31st FG moved to the Maltese island of Gozo. This aircraft of the 308th FS sits in a revetment at Luqa undergoing routine maintenance during August 1943. The fighter's wings are being supported by trestles and the rear panel of the Vokes filter has been removed to allow the groundcrew access to its air cleaner element and engine attachment filter assembly – both areas that required routine cleaning.

First Lieutenant Sylvan 'Sid' Feld was one of the leading, but least known, USAAF aces to fly the Spitfire, and he was regarded as being so ferocious that he sometimes unnerved his own colleagues. Indeed, with nine victories claimed over Tunisia and during the Pantelleria operation, Feld of the 4th FS was the most successful USAAF pilot to fly the Spitfire. After completing his tour with the 52nd FG, he transferred to the P-47-equipped 410th FS/373rd FG in September 1943. Shot down by flak on 13 August 1944 supporting the breakout from the D-Day beaches in Normandy, he was captured and subsequently wounded when caught in a USAAF bombing attack one week later. Feld died of his wounds on 21 August.

'Sid' Feld's Spitfire VC ES276 WD-D had been decorated with his final tally (two victories being painted on the access door) by the time it was photographed in late June 1943. The 4th FS also painted the pilot's name in a scripted style forward of the cockpit. Interestingly, unlike most USAAF Spitfires, this combat-weary machine was marked up in the Type B desert colour scheme. ES276 was later used by the 309th FS and eventually returned to the RAF to be scrapped in 1946.

Parked on the Sicilian airfield of Licata shortly after the invasion of the Mediterranean island in July 1943, Spitfire VC ES317 MX-F of the 307th FS was the regular aircraft of First Lieutenant Ron Brown.

During the spring of 1943 quantities of the improved Spitfire IX began being delivered to the USAAF in North Africa. Among those sent to the 4th FS was EN354 WD-W, christened *Doris June II*, which was the aircraft of First Lieutenant Leonard V. Helton. He claimed two victories in April and August 1943. Photographed at La Sebala in June 1943, this aeroplane featured USAAF national markings on the fuselage, upper wing surfaces and the underside of the starboard wing, but retained an RAF roundel beneath the port wing.

With its RAF wing and fin markings still in place, 4th FS Spitfire VC JG878 WD-V undergoes engine runs after reassembly, probably at Bocca di Falco, in Sicily, in October 1943. The squadron was engaged in coastal escort operations over the Mediterranean at the time.

It was not all sun in the desert, as evidenced by Spitfire VC VF-E of the 5th FS following a flash flood at La Sebala in the summer of 1943 caused by a period of torrential rain. From 28 June that year USAAF aircraft were required to have white rectangles, with an insignia red border, applied to either side of the existing cockades.

Future Mustang ace Second Lieutenant Dan Zoerb of the 2nd FS/52nd FG saw plenty of action in the Spitfire during 1943. He was lucky to make it back to base in one piece after his Spitfire VC QP-A suffered several flak hits during a strafing mission over Italy. At the very least, this aeroplane would have required a new starboard wing and tailplane. Note that the starboard 20mm Oerlikon cannon have been removed in anticipation of the wing being replaced. QP-A wears the short-lived 'star and bars' outline in insignia red.

Spitfire VC JK777 QP-Z of the 2nd FS joined the unit in July 1943, and it is seen here at Bocca di Falco, from where the squadron protected the Sicilian port of Palermo. The aircraft was flown regularly on such patrols during August and September by future ace Second Lieutenant 'Dixie' Alexander. Both aeroplanes appear to have had the insignia red outline to their national markings replaced with insignia blue as per the USAAF directive of 14 August 1943. This change was implemented for the benefit of American aircraft operating in the Pacific.

The 31st FG's final CO during its Spitfire phase was Col C.M. 'Sandy' McCorkle, who took over the unit from Lieutenant Colonel Frank Hill in late August 1943. McCorkle, who would subsequently claim five Spitfire and six Mustang victories, initially flew this personally coded, and relatively clean, clipped wing Mk VB that featured insignia red outlined 'stars and bars'.

Over Salerno on the day of the Allied landings in southern Italy (9 September 1943), First Lieutenant Carroll Pryblo was hit by 'friendly' anti-aircraft fire and forced to crash-land Spitfire VC JK707 on one of the invasion beachheads. He survived the experience and eventually mustered three victories, two of which were claimed in Spitfires.

On 9 April 1943, First Lieutenant Fred Ohr (the only American ace of Korean ancestry), who hailed from Oregon, was participating in a 52nd FG sweep over Tunisia when his formation encountered a large number of unescorted Ju 88s. In a one-sided fight, six of the Junkers bombers fell to the guns of the American Spitfire Vs, including one to Ohr, who claimed his first, and only, victory in the British fighter. Serving with the 2nd FS throughout its tour of combat in the Mediterranean, he often flew this Spitfire IX (painted in the RAF's high altitude fighter scheme) on local patrols and ground attack missions when based in Sicily. Ohr achieved ace status during the summer of 1944 when he claimed five more victories following his conversion to the P-51.

Spitfire IX EN329 FM-D was the final Spitfire assigned to 31st FG CO, Lieutenant Colonel Fred M. Dean. Among the first Mk IXs issued to the group, the fighter was kept relatively clean during its ownership by Lieutenant Colonel Dean. The aeroplane later returned to RAF service and was issued to Italy-based No 145 Sqn.

Lieutenant Colonel Dean's replacement, Lieutenant Colonel Frank Hill, is seen here sitting on the trailing edge of the starboard wing of Spitfire VIII JF452 FA-H at Milazzo, in Sicily, in early September 1943. On 13 July 1943, two days after making his final claim, Hill had been promoted to command the 31st FG. He initially inherited Spitfire IX EN329 from Lieutenant Colonel Dean, who had adopted the RAF practice of senior officers personalising their aircraft by identifying them with their initials, rather than unit codes. Hill continued the practice, and the Spitfire IX he took over eventually carried his initials. By early September he was flying superb new Spitfire VIII JF452, which bore his initials, his substantial score and both his name and his wife's. This aircraft, which was camouflaged in standard desert colours, was flown by Hill for the latter part of his time as CO of the 31st FG, which he led throughout the Sicilian campaign, although he made no further claims with it. He passed the aircraft on to his successor, Colonel Charles McCorkle, who assumed leadership of the 31st FG from Hill in mid-September 1943, and he also inherited the latter's Spitfire VIII JF452. McCorkle claimed his first victory (an Me 210) on 30 September, and on 6 January he almost certainly used this aircraft to down his third (an Fw 190). A month later, again in his personal Mk VIII, he claimed his fifth and last Spitfire victory.

After service with the RAF's No 133 'Eagle' Sqn, Second Lieutenant 'Dixie' Alexander transferred to the USAAF and joined the 2nd FS. He enjoyed considerable success flying the Spitfire with this unit until it re-equipped with P-51s, on which he became an ace in May 1944.

The maintenance board for the 308th FS in early September 1943 shows the mixture of Mk VBs, VCs and VIIIs assigned to the unit at this time. Even their engine numbers are noted. Also listed is Lieutenant Colonel Hill's Mk VIII, JF452.

Spitfire VIII JF400 undergoes acceptance tests and gun calibration at Montecorvino airfield prior to joining the 308th FS on operations over southern Italy in the autumn of 1943. The aircraft is still marked up with old-style national insignia.

Second Lieutenant 'Dixie' Alexander of the 2nd FS flew Spitfire VC QP-A from Borgo, on Corsica, during the final weeks of 1943, the aeroplane boasting a spectacular sharksmouth marking. Alexander recorded its demise in his logbook on 21 January 1944: 'Unhappy day – fini QP-A. Recco Piombino to La Spezia. Skip-bombed electric plant, shot up Macchi 200, two trucks, cleaned out one gun post. Was hit by three cannon bursts and lots of machine gun. Flaps gone, one elevator, one wing tip, crashed base, Category C. Bumped head, sore arm and leg.'

Major Garth Jared assumed command of the 309th FS on 9 November 1943 and retained this position until he was shot down and killed by flak over Udine on 18 April 1944 during one of the first missions flown with the Mustang. He had claimed two Spitfire victories prior to his demise.

After having flown some twenty-five combat missions, future ace Second Lieutenant George Loving was allocated this aircraft, which he described as 'a well-worn Mk V', in late December 1943. In honour of his girlfriend, he named it *Ginger* and had the name painted on the nose. The aircraft was also unusual in having dark (possibly olive) green upper surfaces. When over Cassino a few days after it was allocated to him, a fragment of an 88mm flak shell shattered *Ginger*'s windscreen, but caused no other damage. Loving continued to fly this aircraft regularly during the early part of 1944, including covering patrols over the Anzio beachhead, although such missions were increasingly performed by Spitfire IXs. He may have been flying WZ-S when he was involved in a combat with six Bf 109s during a bomber escort mission in the final weeks of the 31st FG's service with the Spitfire, sharing in the destruction of one of them. Loving converted to the P-51 soon afterwards, and 'made ace' with five victories in the American fighter. The original 'Ginger' remained a part of Loving's life, however, as he married her shortly after returning to the US!

This Spitfire IX of the 309th FS/31st FG, named *Eleanor*, was Major Garth Jared's aircraft during his time as CO of the squadron. It displayed his two victory symbols beneath the cockpit, as well as his initials on the fuselage.

Photographed at Castel Volturno airfield in February 1944, Spitfire IX EP615 of the 309th FS/31st FG was assigned to Second Lieutenant Robert Belmont – seen here glancing towards the camera while filling in maintenance logs for the aeroplane. Belmont christened the aeroplane *Thurla Mae III*, and both of his victory claims were applied immediately beneath the crew data block visible under the windscreen.

The very last USAAF pilot to 'make ace' on the Spitfire was New Yorker First Lieutenant Richard Hurd of the 308th FS, who achieved this distinction on 21 March when he claimed two 'Me 109Es' near Pignataro while flying a Mk VIII eight days before the type's withdrawal from service with the 31st FG.

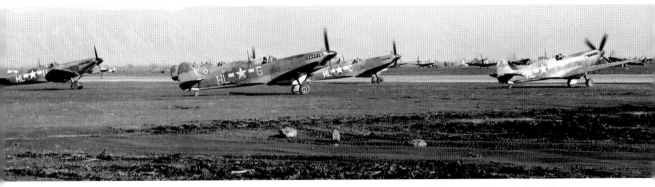

Spitfire VIIIs and IXs from the 308th FS (plus a single machine with an RAF roundel partially visible behind HL-G – probably a newly delivered attrition replacement) taxi out at Castel Volturno at the start of one of the squadron's last missions with the Supermarine fighter in March 1943. The aircraft in the foreground all appear to be fitted with a single 45-imperial gallon 'slipper' drop tank on the centreline. These were available in 30-, 45- and 170-gallon capacity, with the latter giving the aeroplane an impressive 1,500-mile range. When carrying the filled 45- or 170-gallon tank the aircraft was restricted, once airborne and at cruising altitude, to straight and level flight.

Delivered to the 309th FS at Castel Volturno in January 1944, Spitfire IX MH894 was christened *Lady Ellen III* by its pilot, First Lieutenant John Fawcett. The aircraft, camouflaged in the RAF grey/green/grey temperate scheme, was photographed by Fawcett parked on pierced steel planking (PSP) at the airfield between missions soon after it has been assigned to the 31st FG. The aircraft has sandbags for chocks and is fitted with a small 30-imperial gallon slipper tank.

First Lieutenant John Fawcett smiles for the camera while sat in the cockpit of *Lady Ellen III*. The aeroplane was named after Fawcett's wife, and like other fighters in the 31st FG, it had two repeated code letters (JJ) to distinguish it from other older aircraft in the 309th coded J.

For this photograph Fawcett climbed out of the cockpit on the port wing, squatting down beside the aircraft's name. MH894 was subsequently supplied to No 326 'French' Sqn following its replacement by a P-51B in the spring of 1944, the fighter being transferred to the *Armée de l'Air* in November 1945.

First Lieutenant Leland 'Tommy' Molland (left), who claimed 4½ of his 10½ victories in the Spitfire, relaxes with his 308th FS CO, Major James Thorsen, who was credited with two kills, including one in the British fighter.

The 308th FS's First Lieutenant Molland hailed from North Dakota, and christened his allocated Spitfire VIII *Fargo Express* after the stagecoach company of the same name. It is seen here after being marked with his fifth victory, which he achieved on 22 February 1944.

One of 'Dixie' Alexander's last Spitfire VCs is seen here after its withdrawal from service with the 2nd FS in March 1944. This machine is believed to be the aircraft in which he made his final Spitfire claims in February 1944. Behind it is Mk VC JG883 from the 309th FS, which appears to have had its upper surfaces repainted in dark (possibly olive) green.

2nd FS CO Major Bert Sanborn had reason to be grateful for the RAF's antiquated Walrus amphibious biplane when, on 11 April 1944, just before the 52nd FG replaced the last of its Spitfires, his aircraft was hit by flak and he was forced to bail out into the sea. Among those who flew top cover for him were future aces 'Dixie' Alexander and Second Lieutenant 'Sully' Varnell.

When shot down, Sandborn was described as 'flying his uniquely light-coloured aircraft', which may have resembled this unusually camouflaged Spitfire IX of the 4th FS, seen in early 1944 plugged into its trolley accumulator at La Sebala airfield in preparation for another sortie. According to the photographer, Captain John Blythe of the 4th FS, this aeroplane was a 'mottled light brown on the top blended into a light sky blue on the underside'. The pilot manning this 2nd FS machine is First Lieutenant James Puffer, with his crew chief ready to spring into action under the port wing. Blythe also recalled that: 'Our policy at that time was to have two pilots strapped into their aircraft ready to take-off at the signal of a red flare. Six more pilots were in our ready tent with a jeep available.'

On 30 March 1944 the 309th FS officially exchanged its Spitfires for Mustangs at Castel Volturno. Most of the Supermarine fighters seen in this photograph exhibit relatively fresh factory-applied RAF temperate schemes, although the aeroplane nearest the camera is an early production Mk IX in a very faded desert scheme. Coded VV, it was flown by the squadron CO, Major Garth Jared. The Spitfires next to this aircraft, RR and XX, were named *Janice* and *Gaye*, respectively.

All three squadrons within the 31st FG kept a handful of Spitfires as 'hacks' after transitioning to the Mustang, including this machine used by the 307th FS. Painted silver overall, the aeroplane was finished off with the red tail stripes adopted by the 31st FG as its group marking. Note that the fighter also has 'toned down' white areas within its fuselage 'star and bar'.

This was the fate of many veteran Spitfires that survived the conflict in the MTO. Languishing in a scrapyard somewhere in Italy immediately post-war, these Spitfire VBs and VCs exhibit a mix of USAAF and RAF national markings. The HB-coded aircraft in the foreground had previously been flown by No 229 Sqn, which left its aeroplanes behind when the unit was transferred from Sicily to the UK in April 1944.

Chapter Three

Training and Photo-Reconnaissance Units

Although both the 31st and 52nd FGs had departed British shores as long ago as late October 1942, the groups' original Spitfire VBs had remained very much in the UK. Some examples were sent to the 4th FG, others reverted back to RAF control and still more were passed on to the recently arrived 67th Observation Group (OG), which was assigned to the Eighth Air Force in late September 1942. Although a dedicated tactical reconnaissance unit, the group initially flew unmodified Spitfire VBs to allow pilots to become familiar with their unique low-level missions in the ETO. The 67th OG controlled no fewer than four squadrons at Membury, in Wiltshire, and this former 4th FS/52nd FG aircraft was serving with one of these units when it force-landed near the airfield on 13 February 1943. Having suffered only modest damage while coming to a halt in a freshly ploughed field, the aircraft was undoubtedly returned to service in fairly short order.

By the time this photograph of Spitfire VB BM181 was taken at Membury in the summer of 1943, the 67th OG had been redesignated the 67th Reconnaissance Group (RG) and all USAAF Spitfires in the ETO had had their national insignia modified to include white horizontal bars outlined in insignia red. BM181, assigned to the 107th Reconnaissance Squadron (RS), was used extensively in the training of USAAF pilots destined to fly photo-reconnaissance missions later that year. Issued new to Fighter Command's Nos 457 'Australian' Sqn on 30 April 1942, BM181 was transferred to No 611 Sqn at the end of the following month and then passed on to No 81 Sqn on 30 July. The Spitfire eventually joined the 67th OG in June 1943, remaining with the unit until February 1944 when it was passed on to Cunliffe–Owen for conversion into a Seafire IB. The fighter later served with 761 NAS at Henstridge, in Somerset.

Well-worn Spitfire VB W3815 was assigned to the 496th Fighter Training Group, which was formed at Goxhill, in north Lincolnshire, in December 1943. Of the three units within the group, only the 555th FTS flew Spitfires, operating a handful of Mk VBs alongside P-51Bs – the latter arrived from February 1944 when the 496th moved to Halesworth, in Suffolk. Clipped wing W3815 was a presentation Spitfire christened *Sierra Leone II*, the aircraft having been paid for with a donation of £5,000 by the Sierra Leone Fighter Fund. Delivered new to No 611 Sqn in late August 1941, the aeroplane saw combat from Hornchurch, in Essex, with this unit and with No 64 Sqn from November of that same year. In early August W3815 joined the 4th FS/52nd FG, moving on to the 109th OS/67th OG three months later. Returning to RAF service twelve months later, it was issued to No 332 'Norwegian' Sqn in November 1943. Reassigned to the 67th Tactical Reconnaissance Group (TRG) in September 1944, it was one of a number of elderly Spitfire VBs passed on to the 555th TFS shortly thereafter when the poor serviceability rates of its battle-weary Mustangs began to have an adverse impact on the unit's effectiveness. Rejoining the RAF following a landing accident (the fourth it had experienced it a long operational career), W3815 survived the war as a ground instructional airframe with No 4 School of Technical Training at St Athan, in Wales.

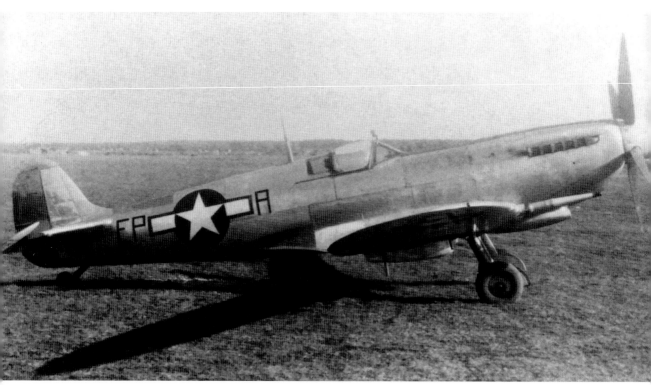

Spitfire VB W3364 was another veteran machine that had seen RAF service prior to being issued to the USAAF. Delivered new to No 602 Sqn in June 1941, the fighter subsequently served with Nos 602 and 81 Sqns later that same year. Transferred to the 2nd FS/52nd FG in early August 1942, W3364 was passed on to the 82nd FG at Eglinton three months later. Handed back to the RAF in early 1943 after the 82nd was sent to fly P-38s in North Africa, the fighter was briefly assigned to No 52 OTU prior to returning to USAAF control. Allocated to the 153rd TRS/67th TRG, the aeroplane was stripped of its camouflage and adorned with non-standard codes inspired by the initials of its assigned pilot, Captain Emmette P. Allen. Note that the fighter has also been fitted with a four-bladed propeller, possibly sourced from one of the unit's Spitfire PR XIs.

Stripped of their armament and wing tips, these Spitfire VBs (BL680 is furthest from the camera) of the 7th Photographic Reconnaissance Group were photographed at Goxhill in late 1943. Both were assigned to the 14th PRS, which was the only unit within the 7th PRG to operate Spitfires through to April 1945 – it also had F-5 Lightnings, which equipped the group's remaining three squadrons. Part of the Eighth Air Force's 325th Photographic Wing, the 7th was established at Mount Farm, in Oxfordshire, in June 1943 and initially issued with five unarmed and clipped wing Spitfire VBs for maintenance and flying training. Spitfire PR XIs duly arrived four months later. Note the different propeller spinners fitted to these veteran aircraft.

The tail section of Spitfire VB BL680 is inspected closely by pilots from the 14th PRS at Mount Farm. These individuals would use this machine and a number of other war-weary Mk VBs to build up their experience in the Spitfire prior to graduating to the PR XI, which they would fly on reconnaissance missions over occupied Europe.

With his Spitfire connected up to a trolley accumulator, the pilot of BL680 runs through his final preflight cockpit checks prior to starting the fighter's Merlin 45. Issued new to No 616 Sqn in February 1942, BL680 eventually joined the 14th PRS/7th PRG in November 1943. Unsurprisingly, considering the nature of its employment as a familiarization aircraft, it suffered landing accidents in January and March 1944 before finally being written off when it crashed on take-off from Watton, in Norfolk, on 7 May 1944.

PR XI MB950 was one of the first photo-reconnaissance Spitfires supplied to the 7th PRG, being received by the RAF on 18 October 1943 and transferred to the group on 13 November. The aeroplane is seen here being prepared for an operational mission shortly after its arrival at Mount Farm, the pilot being assisted with his cockpit drills by his crew chief while photo technicians from the 22nd PRS complete their final checks on the fuselage-mounted F24 14in camera mounted for oblique photography in MB950's 'X Type' installation. Immediately beneath the oblique camera were the two vertical F52 20 or 36in cameras in the 'Universal' installation that shot through the underside of the aeroplane. Subsequently nicknamed *Upstairs Maid*, this aeroplane served briefly with the 22nd PRS until replaced by an F-5 in January 1944. MB950 was then transferred to the 14th PRS, flying with this unit through to early October, when it was returned to the RAF.

VCS-7 was the only US Navy squadron to fly the Spitfire, which it used for bombardment spotting during the D-Day landings. The unit was manned by pilots assigned to the SOC Seagull and OS2U Kingfisher floatplane flights embarked in the battleships and cruisers committed to Operation *Overlord*. The floatplanes had been deemed to be too slow and vulnerable to enemy aircraft and ground defences to allow them to carry out their vital spotting role for the naval vessels, which had been tasked with knocking out the German coastal defences on the eve of the landings. In April 1944 the Seagulls and Kingfishers from six ships were flown ashore and placed in open storage and the pilots assigned to the newly formed VCS-7. Trained to fly Spitfire VBs at Middle Wallop, in Hampshire, by pilots from the USAAF's F-6 Mustang-equipped 15th TRS, the naval aviators then took their aeroplanes to Lee-on-Solent. Here, they became part of the Fleet Air Arm's Third Naval Wing, and from 6 June VCS-7 was one of seven units tasked with gunnery spotting over Normandy. Flying in pairs, the VCS-7 pilots completed thirty-four sorties on D-Day. This mission intensity continued for four days, but then steadily decreased until the unit was disbanded on 26 June. Shortly thereafter, the naval aviators flew their floatplanes back to their respective ships. Here, Lieutenant Robert Doyle and Ensign John Mudge congratulate each other after helping break up a German armoured column on 10 June.

Chapter Four

Spitfires in Glorious Colour

Presentation Spitfire VC AA963 *Borough of Southgate* was delivered to the RAF in November 1941 and shipped to the USA for evaluation on board the merchantman SS *Evanger* three months later. Reassembled at Wright Field in Dayton, Ohio, in March 1942, the aeroplane was tested by the USAAC after a brief spell on display in RAF colours at Chicago Municipal Airport. By the time this photograph was taken of the aircraft (minus its main wheels), it had been repainted in a USAAF Olive Drab and Neutral Gray scheme, complete with unusual blotching on the tail surfaces. The fighter was subsequently scrapped in the USA.

Although not technically a 'star-spangled' Spitfire, Mk VB AB875 was assigned to No 71 'Eagle' Sqn at Martlesham Heath, in Suffolk, in late August 1941 – a little more than a year later this unit would become the 334th FS of the 4th FG at Debden. The artwork adorning the nose of the aeroplane was inspired by the No 71 Sqn insignia. The aviator posing on the wing root of the fighter is Pilot Officer Joseph M. 'Moe' Kelly, who volunteered for service when just 19 years old. Hailing from Oakland, California, he was one of four 'Eagles' to later see action flying Kittyhawks with No 450 'Australian' Sqn in North Africa in 1942–43. The quartet of Americans had requested a transfer to the Far East to fight the Japanese in the China–Burma–India theatre, but got no further than Egypt! Delivered new to No 71 Sqn, AB875 was damaged in a heavy landing at Martlesham Heath in March 1942. Once repaired by Air Service Training Ltd, it was passed on to No 308 'Polish' Sqn in early March 1943. AB875 was transferred to No 350 'Belgian' Sqn seven months later and to No 322 'Dutch' Sqn in January 1944. Damaged in yet another accident in March 1944, the veteran fighter saw out the final months of the Second World War with the Central Gunnery School. It was struck off charge in December 1945.

First Lieutenant William Skinner of the 308th FS/31st FG runs his hand over damage inflicted on his Spitfire VC by an 88mm flak shell that detonated near his aircraft during a mission over Italy in October 1943. Note the recently applied white bars (minus their insignia red border) flanking the national insignia.

First Lieutenant William Skinner replaced his war-weary Spitfire VC with this Mk VIII, which he christened *Lonesome Polecat*. The fighter is seen here at Castel Volturno in February 1944 with Skinner's armourer and crew chief perched on the wing. Note the faded state of the aeroplane's dark earth and mid stone camouflage scheme.

As noted in the previous chapter, PR XI MB950 was one of the the first photo-reconnaissance Spitfires supplied to the 7th PRG in November 1943. All PR Spitfires assigned to the USAAF were supplied in standard RAF PRU blue overall, with 30in fuselage and wing 'stars and bars'.

During 1944 MB950 was adorned with a horizontal red stripe on the cowling, which was a 7th PRG marking applied to all its aircraft. Seen here landing back at Mount Farm following the completion of another long-range PR flight over occupied Europe, MB950 also boasts an all-green rudder – the squadron colour for the 14th PRS.

Both wearing 'WW' letters on their tails, as well as their RAF serials, these clipped wing Spitfire VCs were used by the 14th PRS as 'hacks' during the spring and summer of 1944. 'WW' stood for War Weary, which meant that they were unsuitable for combat and fit for training use only. EN904 in the foreground shows signs of heavy weathering, while AR404 has been stripped of its RAF camouflage and black anti-glare paint applied immediately forward of the cockpit.

Originally delivered to the RAF's No 124 Sqn in June 1942, AR404 was passed on to the 52nd FG's 2nd FS two months later. Damaged in a belly landing in October of that year, it was supplied to No 416 'Canadian' Sqn following repairs and conversion into an LF VB (clipped wings and re-engined with a Merlin 45B) in April 1943. Transferred to No 313 'Czech' Sqn five months later, AR404 was supplied to the 14th PRS/7th PRG in November of that year. Enduring yet another belly landing shortly thereafter, the aeroplane was stripped of its armament and declared a 'hack' for local area orientation flights by would-be Spitfire PR XI pilots. Returned to the RAF in September 1944, the fighter was sold to the Portuguese Air Force in 1947.

The dark green and ocean grey upper surfaces on EN904 have weathered to such an extent that the aeroplane almost looks like it has been camouflaged in a two-tone blue scheme. Delivered new to No 416 'Canadian' Sqn at the end of June 1942, this aircraft was transferred to No 602 Sqn two weeks later. In September of that year it joined No 164 Sqn, before being sent to No 341 'Free French' Sqn in February 1943. Two months later it was assigned to No 340 'Free French' Sqn, and from there EN904 joined the 14th PRS/7th PRG at Mount Farm in mid-November 1943. It too suffered an early landing accident the following month, after which it was declared War Weary and used exclusively as a 'hack'. Sustaining Category E damage on 26 March 1945, the fighter was struck off charge the following day and scrapped.

Featuring recently applied, reduced, half-D-Day stripes, these PR XIs will soon be refuelled with 100 octane gasoline from the bowser that has just arrived at their Mount Farm dispersal behind a 4½-ton Autocar U7144T tractor unit. Both Spitfires (PL914 to right) wear yellow flight spinners, applied in August 1944.

Tanks full, PL914 and the unidentified PR XI next to it have their Merlin 70 engines run up prior to the pilots taxiing out at the start of another photo-recce mission in October 1944. The aeroplane with its engine and cockpit covered with tarpaulins is PA841, nicknamed *Kisty the 1st*, which was among the first PR XIs supplied to the 7th PRG in late 1943. A veteran of many sorties with the 14th PRS, the aircraft was damaged in a take-off accident in August 1944. Repaired, PA841 remained with the unit through to war's end. Note the white-spinnered PR XIs parked in front of the blister hangar behind PA841.

PL914 was supplied to the 7th PRG as an attrition replacement in early October 1944, serving with the 14th PRS until it was returned to RAF service in early April 1945. Note the red paint markers overlapping the tyre and wheel hub, which were applied in this way to indicate if the tyre had slipped around the wheel rim. The deepened nose cowling for the PR XI's enlarged oil tank and the frameless windscreen (a feature of all PR Spitfires) are clearly visible in this view, as is the aeroplane's weaponless port wing.

MB946 was another very early PR Spitfire arrival at Mount Farm, this aeroplane joining the 7th PRG's 13th PRS in November 1943. It was one of a handful of recce Spitfires to boast a mission tally, which took the form of small red swastikas applied in rows just forward of the cockpit. Transferred to the 14th PRS shortly after joining the group, the aeroplane was eventually written off after suffering an accident in late March 1945. Attached to a trolley accumulator for ground power, it is seen here in the process of being refuelled between sorties. Compared with early versions of the Spitfire, which held just 85 gallons of internal fuel, the PR XI's enlarged tanks could hold 218 gallons. This gave the aeroplane a range of 1,360 miles – with an external drop tank fitted, a round trip to Berlin was possible.

PR XI PA944 also featured mission markings forward of the cockpit. Transferred from the RAF to the 7th PRG in April 1944, the aeroplane was written off in a landing accident at Mount Farm following a mission to Germany on 12 September that same year. Its pilot, Captain John S. Blyth, somehow managed to get his headset cord caught up around the activation handle for the aircraft's CO_2 undercarriage deployment system, locking the landing gear in the up position. This in turn meant that he had to perform a belly landing upon his return to base. Sent to Heston Aircraft Ltd for repairs, the aeroplane was still there on VE Day and was duly struck off charge.

Assigned to the 7th PRG in mid-January 1944, PA892 was initially nicknamed *High Lady*. Among its more notable mission was a sortie to Berlin on 6 March 1944 with Major Walter Weitner at the controls – he damaged the aeroplane in a landing accident at Mount Farm eighteen days later. Once repaired, it returned to the group and was christened *My Darling Dorothy*, as seen here in this October 1944 photograph that was taken shortly after the Spitfire's D-Day stripes had been crudely painted over with PRU blue. Note also the overpainting of the national insignia's dark blue border. PA892 was returned to the RAF in early April 1945 and subsequently scrapped.